Study and Communication Skills for the

CHEMICAL SCIENCES

Tina Overton, Stuart Johnson, and Jon Scott

OXFORD UNIVERSITY PRESS

OXFORD
UNIVERSITY PRESS

Great Clarendon Street, Oxford OX2 6DP

Oxford University Press is a department of the University of Oxford.
It furthers the University's objective of excellence in research, scholarship,
and education by publishing worldwide in

Oxford New York

Auckland Cape Town Dar es Salaam Hong Kong Karachi
Kuala Lumpur Madrid Melbourne Mexico City Nairobi
New Delhi Shanghai Taipei Toronto

With offices in

Argentina Austria Brazil Chile Czech Republic France Greece
Guatemala Hungary Italy Japan Poland Portugal Singapore
South Korea Switzerland Thailand Turkey Ukraine Vietnam

Published in the United States
by Oxford University Press Inc., New York

British Library Cataloguing in Publication Data

Data available

Library of Congress Cataloging in Publication Data

Data available

Typeset by Glyph International, Bangalore, India
Printed in Great Britain by Ashford Colour Press Ltd,
Gosport, Hants

ISBN 978-0-19-953968-0

3 5 7 9 10 8 6 4 2

Preface

The transition from school or college to university is a very significant life event: for all students it marks the beginning of the next stage in their educational career and, for many, it may be their first experience of living away from home for any length of time. This transition, therefore, is associated with a whole range of new experiences in social and academic terms. Among these is a marked change in learning and teaching styles, with an expectation of increasing independence as a learner and development of a more mature, critical approach to the subject. Meeting the challenges posed by these changes requires the development of new skills in many areas of life, not least in the areas of study and communication. If you can develop effective study and communication skills early in your university career, you will facilitate your overall learning and help maximize your academic performance.

This book arose from many years of experience in developing and delivering courses in study skills, communication skills, and career planning in the Department of Chemistry at the University of Hull and in the School of Biological Sciences at the University of Leicester. The structure of the book and its format are based on our experiences of delivering these courses, in terms of recognizing where students experience difficulty in developing these skills and in learning from those students better ways of teaching and providing support for their learning. We hope that the future generations of students will be able to benefit from this guidance and we also look forward to learning more about the ways in which skills can best be developed.

T.L.O.
University of Hull
S.J. and J.S.
University of Leicester

Acknowledgements

We would like to thank the students in the Department of Chemistry at the University of Hull and in the School of Biological Sciences at the University of Leicester who have helped to shape our work and, over the years, have experienced the evolution of our courses in Study and Communication Skills.

Our thanks also go to Laura Quigley and Jonathan Crowe and Oxford University Press for their support throughout the project and for not harassing us too much when our timings slipped.

Contents

Chapter 1

Why are study and communication skills important?

Introduction

As the title suggests, this book explores the key study and communication skills that are needed by chemistry students beyond school level. This means if you're studying, or about to study, chemistry this book can help. As you may have already gleaned from the contents pages, this book addresses a range of important topics, including:

- making the most of lectures;
- getting the most out of tutorials and group work;
- maximizing the skills developed through practical work;
- working with different information sources;
- choosing the right writing style;
- writing essays;
- writing practical reports;
- working in tutorials and groups;
- avoiding plagiarism;
- preparing scientific presentations;
- delivering scientific presentations;
- creating academic posters;
- using feedback;
- getting the most out of revision;
- getting the most out of exams;
- career planning;
- getting your first graduate position.

We have written this book to help you in all of these areas and you may wish to work through them one by one. Alternatively, however, it may be that you have a specific issue that you need help with and so want to skip directly to the relevant chapter. Either approach is fine: we have written the book so that it makes sense when read as a whole text or as individual chapters. It is a good idea to read this chapter first though, as it helps set the scene.

Good study and communication skills are vital if you are going to do well at university; they will help you to study efficiently and effectively and so perform to the best of your ability in your chosen area of chemistry. However, good study and communication skills aren't only useful for studying for your course; the kinds of skills you need to develop to do well in your chosen field of study have many parallels with the kinds of skills that will be useful throughout your career. Being able to research a subject, construct an argument, write a report, present information, manage your time, and plan your own development are all skills that are highly sought after by most, if not all, employers. So we hope this book won't just help you to do better in your degree, but will also help you to do well beyond university.

Incidentally, in this chapter we mainly talk about studying for a degree at university. However, this book is equally useful if you're doing another kind of chemistry course, whether at university or college.

1.1 'I did study skills at school. Why do I still need to read this book?'

Clearly, if you have got as far as getting a place at university you must be good at studying. Also, you were probably taught about study skills at school or college, and you will have certainly developed your own study skills during your time there. So why do you need to read this book? The main reason is this: studying for a degree at university is different from studying at pre-university level. There are numerous ways in which it's different. The important ones are as follows:

- studying at university requires you to be more independent than you were at school. You will therefore be expected to undertake much more self-directed study and so will need to learn to **manage your time well**.
- studying at university requires you to be more critical about your subject than you were at school, and not just accept everything at face value without question. You will be expected to identify differing views on a topic and develop your own views too. So you will need to learn to **think critically**.
- studying at university requires you to be more self-reflective about your ability and performance than you were at school in order to adapt and develop your current skills to better suit an undergraduate level of study. So you will need to learn to **develop yourself**.

Studying at university is, for most, a more challenging experience than school or college. This book will help you to meet those challenges and assist you in making the transition to become an accomplished undergraduate student. This chapter will focus in on the above differences to highlight what we consider to be the foundational (or fundamental) skills you need to develop (and keep in mind) throughout your studies.

1.2 The contextual nature of skills: making the transition to undergraduate study

Skills are contextual. By this we mean that skills are learnt in a particular context and the way you apply them is specific to that context. For example, when you learn to drive a car you learn in a certain way—usually with a driving instructor sitting next to you, in a car that's familiar to you, and receiving instruction on what to do. But just because someone has learnt to drive, it doesn't mean the person is capable of driving well in any and every situation. The chances are, initially at least, that the driving skills will be limited to a similar context to that within which they were originally learnt. Read through the following scenario to see what we mean.

Imagine a friend of yours has spent the last few months learning to drive and they have just passed their test, passing with only two minor faults.

- If they offered you a lift, in a similar car to the one they had learnt in, would you feel safe catching a lift with them?
- If they offered you a lift, in a similar car to the one they had learnt in, but it was night time, would you feel safe catching a lift with them?
- If they offered you a lift, in a similar car to the one they had learnt in, but they were going to drive on a busy motorway, would you feel safe catching a lift with them?
- If they offered you a lift, but they were driving an articulated lorry, would you feel safe catching a lift with them?

The chances are you would probably be more reluctant to accept a lift the more unusual the context became. This is because the more unfamiliar a context is to someone the more difficult it is to perform well in that context.

There are clear parallels between this example and making the transition to undergraduate study which we are sure you will have spotted. You already have study and communication skills, probably ones that have served you well so far, but as we have seen, studying for a degree at university is different from studying at pre-university level. Your skills will therefore need adapting and developing to fit the new situation you are encountering. The good news is that skills can be transferred from one context to another, so the skills that you have developed so far are by no means wasted. In fact, they provide a very useful starting point.

1.3 Foundational skills

As we mentioned above, the important differences between studying at pre-university and at undergraduate level reveal three foundational skills that are vital for students who want to perform well during their degree: managing your time, thinking critically, and developing yourself. These aren't the only skills—and subsequent chapters highlight many others—but they do underpin the ones that follow and so are worth drawing your attention to at this early stage.

1.3.1 Manage your time well

Many students, on nearing the end of their degree course, wish that they could have their time again. This is often because they feel that, given another go, they could do better. The reasons for thinking they could do better may be quite diverse, but common to many of them will be the notion that if they had managed their time more effectively they would have been able to perform more effectively, and get a better degree. Therefore, managing your time well is crucial to performing well at university. There will be many demands on your time—social as well as academic—and you will have more autonomy about how you choose to spend your time compared with school or college. It's important, therefore, that you learn to allocate your time appropriately. This is one of those things that is easier to say than to actually do; in theory it's straightforward, in practice it's difficult.

We have identified the need to manage your time well as a foundational skill because it is necessary for performing well in a wide range of academic tasks, but that doesn't mean we are going to deal with it in an abstract way—with a chapter just about time management, for example. Instead we address it in what we think is a much more helpful way: giving you advice about how to manage your time wherever it is relevant, in context—for example, when writing an essay or assignment, giving a presentation, or preparing for exams. As such, effective time management is a theme that occurs throughout the book, so look out for these sections in the chapters that follow.

1.3.2 Think critically

The second foundational skill that we want to draw your attention to in this opening chapter is the need to learn to think critically. Critical thinking is common to all academic endeavours and so it is important to understand what it is. Many people think about 'critical' as a negative term, and perhaps words like 'unfavourable', 'fault-finding', or even 'unkind' spring to mind. Being critical is, however, much more than this. If you have ever read a review of a film, book, play, or concert, the review will have been written by a 'critic': someone whose professional job is to be critical in its full sense. For example, the critique of the performance of a play comments on all sorts of aspects of that play, including the quality of the performances of the individual actors, the way the stage was designed, the lighting, and so on. Thus, the critic may have written in glowing terms about the leading lady but felt that the supporting cast was not up to the job. In this context being critical may mean being very positive and negative simultaneously, about different aspects of the same thing. However, it is important to note that this kind of criticism is subjective: one critic's view of the play may be very different from another's.

In the academic sense, the word 'critical' also doesn't have specifically negative connotations. In fact, in an academic context, being critical is a good thing and is something to be encouraged. However, there is one key difference between thinking critically in scientific terms and in terms of reviewing a play: in science, the critical thinking should be undertaken *objectively*. For example, if you are comparing two theories, you should compare the ideas being presented and evaluate the weight of evidence supporting them, using this evaluation to decide which is most tenable. You should approach this evaluation from a neutral (impartial) viewpoint: not looking for something to be right or wrong because you *think* it should be, but because the available evidence indicates it to be so. Critical thinking is therefore a process that requires the evaluation of evidence to come to a conclusion that is supported by the outcome of experiments or observations: it is an intellectual exercise that is fundamental to the way in which science works and develops: the basis of scientific method.

As with managing your time, this is not a book primarily about thinking critically. Rather we will address what it means to think critically when undertaking particular academic tasks—for example, being involved in lectures or tutorials, or when researching information for an essay or presentation. Look out for these sections in the chapters that follow.

1.3.3 Develop yourself

The third and final foundational skill that we have identified is the need to learn to develop yourself. This won't be an unusual concept to you as doubtless you will have been encouraged to engage in some kind of personal development planning at school. To make the transition from pre-university to undergraduate level, however, you will need to continue to develop yourself. Most universities will have a personal development planning scheme that you will be encouraged to take part in. Such schemes are designed to help you in two ways: to improve your academic performance, and to help you make plans for what you will do once you have graduated. It is a process of continuous improvement: thinking about and reflecting on what you have done in the past, and learning from this experience to influence in a positive way what you do in the future (asking yourself questions like 'what worked well?', 'what didn't go as planned?', 'how can I make sure that the things that didn't work well work better next time?').

Regardless of the format of the system you use, the important thing is that you reflect on your progress (to identify where you are doing well and where you need to improve) and make plans for your future development. You will have numerous opportunities for reflection, including:

- feedback on coursework;
- feedback on exams;

- conversations with your lecturers;
- conversations with your personal tutor;
- conversations with friends;
- time by yourself thinking.

It is important that you use these opportunities to assess how you are doing and then (just as importantly) decide what you are going to do about it. If you have done well on something, then celebrate it in some way (you probably don't need much encouragement to do this!) but also make sure you are aware of why you got a good mark so you can take that approach again. Equally, when you identify areas for improvement, plan what you need to do to improve. Reading this book is a good start.

1.4 Checklist

Here is some brief advice for you.

Don't:

- expect studying at university to be the same as at school or college;
- expect your tutors to 'spoon feed' you every piece of information that you need to know.

Do:

- take responsibility of your own learning;
- learn to manage your time effectively;
- get into the habit of reflecting on your own progress.

* Chapter summary

Good study and communication skills are vital if you are going to do well at university. Undergraduate study is different from pre-university study in a number of important ways. The skills you have already gained up to this point will be useful, but you will need to develop and adapt them for the new context you now find yourself in. Managing your time well, thinking critically, and being committed to self development underpin the other skills that we will now explore in the subsequent chapters.

Chapter 2

Making the most of lectures

➲ Introduction

Lectures are a prominent and important feature of all undergraduate chemistry courses, therefore being able to make the most of them is an important skill. The lectures you have as part of your course will be many and varied; your lecturers will have different styles and different levels of ability in lecturing, and the lecture content itself will be of more or less interest to you and more or less demanding. Regardless of these variables, however, you need to make the most of lectures. This chapter will help you understand how to make the most of lectures by first addressing the purpose of lectures and then giving advice on how to:

- prepare before lectures;
- listen and make notes during lectures;
- follow up after lectures.

2.1 The purpose of lectures

Lectures can be a very valuable resource as they can synthesize the views of several researchers and textbooks or provide new, and even unpublished, information. Lectures are often liked by academics as a means of communication because they are a very efficient means (in theory at least) of transferring lots of information to large numbers of students all at once. How efficient a lecture actually is at communicating information will depend on at least two variables:

1. how well the lecturer prepares and delivers the lecture;
2. how well the audience is listening and making notes.

Clearly, you can't do much about the first of these variables (how well the lecturer prepares and delivers the lecture), at least not in the short term. However, do make sure you feedback to staff your thoughts on how you found a particular module, usually via the end of module survey (if you think your feedback needs acting on more urgently though, contact the module convenor directly). The second variable (listening and making notes) is clearly an area you can do something about; in fact, no one else can do it for you!

2.1.1 More than information transfer

Lectures are not merely about transferring information from the lecturer to the audience. In fact, we would argue that they are not even primarily about transferring information from the lecturer to the audience. If the purpose of a lecture was only to transfer information from the lecturer's brain to the students' notes, then lectures really have very little point because there are many more efficient ways of doing this; for instance preparing a handout, or suggesting particular chapters of a core text to read. Lectures are of value because they add something in the communication of information that couldn't be achieved as well by other means. You will know this if you have ever tried to read a handout or a friend's notes from a lecture that you didn't attend; trying to understand the information out of context can often be very difficult. So if lectures aren't primarily about transferring information from the lecturer to the audience, what should they be about? The transfer of information is important; however, it is what you do with the information as you receive it, and afterwards, that really matters. This is where you, as a member of the audience, come in. It is perfectly possible, even sitting in a lecture that has been well prepared and is being well delivered, to disengage your brain and relegate yourself to being a mere passive recipient of the information. As you will see later in the chapter, it is possible to not think about a lecture even while you are taking notes on it. The amount of notes you take isn't the measure of whether you have been paying attention; in fact, you might have a lot of notes because you haven't been paying attention! Doing something useful with the information as you receive it is about processing the information effectively. This processing can't all take place within the lecture itself (even if you do listen actively and make notes appropriately), which is why this chapter covers preparing before lectures and following up after lectures, as well as what to do during lectures. Before we address these issues though, it will be helpful to consider how lectures are different from what you may have experienced at school or college, prior to your undergraduate degree.

2.1.2 Differences between lectures and lessons

Lectures at undergraduate level will be different to lessons you had at school or college in a number of important ways. At the risk of generalizing, these differences probably include the following.

Class size

In the final years at school or college you may well have been in a class of less than 30 students; at undergraduate level it's not unusual, especially in the first year, for class sizes to be up to 200–300 students (although typically, as you move through your course to the final year, the class sizes will get significantly smaller).

Anonymity

A bigger class makes individuals within the class more anonymous; it is easier in a larger class to just sit quietly and not really engage with the lecture content because you feel like no one will notice.

Relationship

Bigger class sizes mean that you are more remote from the lecturer, not only in terms of physical space but also in terms of relationship. The chances are that, in your first year of study, you probably won't get to know the lecturer very well and they probably won't even know your name.

Completeness

In school your lessons probably covered everything you needed to know about your syllabus; at undergraduate level, lectures don't tell you everything you need to know and so need to be supplemented with additional reading (more on this one later).

2.2 Prepare before lectures

You are already on a busy course with a lot of timetabled teaching hours—made up of lectures, tutorials, workshops, and practical classes—is preparing for lectures really necessary? We think it is. Before you discount this as something only for extremely keen and diligent students, bear this in mind: preparing for lectures doesn't have to take long and can actually save you a lot of time. It doesn't have to take long because you are not trying to cover all the material the lecture is going to cover (that's what the lecture is for), instead you are just giving yourself a framework to help you understand it better. It can save you a lot of time because if you have a framework to help you understand the lecture, you can process the information more quickly, thus making the listening easier, the note making more selective, and the follow-up more focused. In order to know how to prepare for lectures though, you first need to know some basic information about your course.

2.2.1 Know your course

In this context, by 'know your course', we mean finding out how your lectures relate to the module as a whole. This will involve finding out answers to questions such as:

- Do the tutorials prepare you for the lectures or do they follow-up on material covered in the lectures?
- Do the practicals prepare you for the lectures or do they follow-up on material covered in the lectures?
- Are there opportunities to discuss the lecture in your tutorials?
- Is there any recommended pre-reading, if so, what is it?
- What are the intended learning outcomes for the module?

Answers to these questions should be available in your module handbook, or alternatively from your tutor. Additionally, you will need to know the title of each lecture, or group of lectures, within any given module; this too should be in the module handbook that may also contain further details about what each lecture will cover.

2.2.2 Reading before the lecture

Re-reading the notes that you took during the last lecture is the easiest way to prepare for a new lecture. This refreshes your memory and enables you to build on prior material most effectively. Other ways of preparing for lectures don't have to take long because you are not trying to cover all the material the lecture is going to cover, but are just giving yourself a framework to help you understand it better. A framework will help you understand the lecture because it gives you somewhere to place information, and so relate it to other information, rather than trying to understand it in isolation (which is always very difficult).

The framework can be quick to create because you only need to understand basic structure and terminology, rather than detail. A good way of achieving this is by looking up the relevant chapter in the core text for the module (this is where you need to know the title of each lecture) and scanning through it to familiarize yourself with the main themes. This doesn't require you to read the whole chapter, or read it in detail, just scan through all or some of the following:

- the chapter overview (if there is one);
- the chapter headings and subheadings;
- the introduction to the chapter;
- the conclusion or summary to the chapter;
- figures, diagrams, or tables (which often summarize a lot of information succinctly);
- also, look out for terminology that you are unfamiliar with and try and find out what it means.

Core texts are not the only source of information for pre-reading, however; your own notes will be useful too. Assuming you are not on the first lecture of a module, your notes from previous lectures in the module will be an important source of information. Again, you don't have to read all the notes in detail, just scan through them to remind yourself of what was covered previously, but when you come to a section that you don't understand, spend more time on it and check up on the topic in a textbook or other reference source, or ask your friends.

If you have prepared for a lecture it makes listening and note making much easier; which are the subjects of our next two sections.

Try this: Prepare for your next lecture session

Re-read your notes from the last lecture. Make a note of any points that are not clear to you. Try to clarify these for yourself by reading the relevant section in the textbook. If you are still unsure make a note to see the lecturer.

Look forward in the textbook to what you will be covering next. Jot down main points, familiarize yourself with the diagrams, figures and new terminology.

2.3 Listen actively during lectures

This section is deliberately titled 'listen actively' because there is a difference between just listening and listening actively. Listening actively suggests that you are alert, attentive and ready to engage with the content of the lecture, as opposed to just being there. Clearly, the quality of the lecture (how well it is prepared and delivered) will have an effect on how easy it is for you to listen, but listening actively means that you are willing to work hard at listening regardless of the quality of the lecture. Additionally, we have already identified that preparing for a lecture makes listening actively much easier, because you have a framework within which to place new information and you will at least recognize the terminology used.

2.3.1 Identify your priority

There will be some lectures when it will be difficult to both listen and make notes, either because of the complexity or newness of information being communicated, or simply the amount. In such situations you will need to identify what your priority is: is it to listen or is it to make notes? Deciding which your priority is will depend on whether you are more concerned with trying to **understand the information** or **collect information**. If your priority is understanding the information then you will need to:

- focus on listening;
- make only brief keyword notes;
- follow-up the lecture by making detailed notes.

If, however, your priority is collecting information, then you will need to:

- focus on making notes;
- make more detailed notes;
- follow-up the lecture by reviewing your understanding of the content.

When making a decision about your priority, also consider what resources are available to you after the lecture. It is usually relatively easy to get hold of appropriate textbooks or perhaps borrow a friend's notes, and you may well also have access to lecture handouts; however, it is much more difficult to actually experience the lecture again. Even if you can listen to a recording it is rarely as good as actually being there. This would suggest that active listening, and therefore understanding, should be more important than making notes. Remember, however, you don't necessarily have to keep to the same strategy all the way through a given lecture; you could switch between the two depending on the material (more on this in Section 2.4).

2.3.2 **Listen for structure**

Making notes is always easier if you have an awareness of the structure of the lecture. Being aware of structure enables you to be more selective, and therefore more focused, in your note making because it gives you an indication of what the important bits are. That's not to say that the other bits are unimportant, it's just a recognition that in terms of understanding there are certain elements of a lecture that are helpful to grasp in order to understand the rest of it.

Sometimes the structure of a lecture is very clear, other times they are less clear. It is important, therefore, to know what the structural clues might be. The most obvious and probably the most common way of making the structure of a lecture clear is for the lecturer to outline what the structure is going to be at the beginning of the lecture, either as a list of headings or as a potted summary of the content. If a lecturer does outline the structure, make sure that you make a note of it straight away (this applies equally whether your priority is focusing on understanding or collecting information), as it will give you a sense of direction and help you to anticipate points or take up the thread of information again should you get lost.

Additionally, during a lecture the lecturer will probably give you some cues, or 'verbal signposts', these include statements such as:

- 'I shall now discuss…'
- 'My next point is…'
- 'Finally…'
- 'In conclusion…'
- 'To summarize…'

These signposts identify a new point or stage in the lecture and you should show this in your notes accordingly. Other signposts include: pausing to indicate a new point or summarizing what has been said prior to moving on.

There are other, more subtle, verbal signposts that can help you structure your notes; you will need to listen for these. Examples include:

- 'On the other hand…'
- 'Others have argued…'
- 'Turning now to…'
- 'Alternatively…'

Other words and phrases indicate that an illustration is about to be given:

- 'An example of this is…'
- 'This can be seen when…'
- 'Evidence for this can be found in…'

Your ability to listen actively, and in particular to listen for structure, will improve with experience. As you improve you will be better able to spot digressions or additional examples and adjust your note making accordingly, which brings us on to the next section.

2.4 Make notes appropriately during lectures

The amount of notes you take isn't the measure of whether you have been paying attention; in fact, you might have a lot of notes because you haven't been paying attention! We therefore need to draw a distinction between making notes and taking notes. We have deliberately referred to making notes so far, as opposed to taking notes, because we think there is an important difference. If you take something you are just the recipient of it, so in the case of lectures you just sit there and write down what the lecturer is saying (or worse, just walk away with the handout).

On the other hand, if you make something you are involved somehow in the creation of it and you bring something of yourself to it, so in the case of lectures you create notes that are unique to you and are produced by you engaging thoughtfully with the content of the lecture. Making notes, therefore, is an active activity, whereas taking notes is usually a passive activity. As you can see, there's quite a difference. That's not to say though that making notes won't involve you writing down word-for-word what a lecturer is saying on occasions; the important thing is that you make notes appropriately. As we have already seen in Section 2.3.1, this will depend a lot on what your priority is; whether it's to **understand the information** (and so you focus on listening) or to **collect information** (and so you focus on note making). In reality though, whilst it is difficult to focus on both listening and making notes at the same time, it is not that you just do one or the other during any given lecture; rather the two go hand-in-hand. You will be constantly switching between listening and making notes all the way through a lecture, but how much you do of each one should reflect your priority.

2.4.1 **Know why you are making notes**

We have already encouraged you to think about what your priority might be when in lectures, and understanding information and collecting information are two key reasons for making notes. However, there are other reasons too. Making notes appropriately in lectures can also help you to:

- concentrate better;
- remember the content better (in the short term at least);
- think about questions you want to ask;
- highlight areas of interest.

Note making, therefore, isn't just about having something to refer to later; it is part of the learning process itself. This is another reason why borrowing a friend's notes is a poor substitute for actually being there and experiencing the lecture for yourself, because so much takes place in the lecture, and in your head, that even really good notes can't really capture the whole event.

2.4.2 **Know how to make notes**

The ability to make good notes is a skill that develops with practice, so don't expect to be an expert at it straight away. Clearly you will have made notes before at school or college, but making notes in lectures in an undergraduate setting is different for a number of reasons, as highlighted in Section 2.1.2 *Differences between lectures and lessons*. So improving your note-making skills comes partly through practice, but there are some important principles to bear in mind too, these include:

- using structure in your notes;
- using your own words;
- using fewer words;
- using abbreviations;
- using space;
- using colour and image;
- using handouts;
- organizing your notes.

We will address each of these principles in turn.

Use structure in your notes

Notes that are lacking in structure will be much more difficult to understand, especially after the lecture, than notes that have a good structure. Imagine you were looking at some

FIGURE 2.1 Poorly structured lecture notes.

TM Chem

d-block, d electrons

Fe [Ar] 4s² 3d⁶

Cr [Ar] 4s¹ 3d⁵

compounds? different

s or d?

Cr(0) [Ar] 3d⁶ ✓

Cr(III) do this?

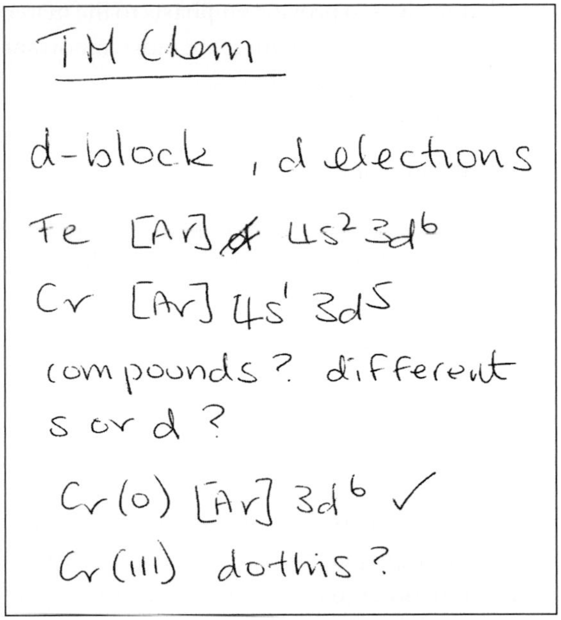

lecture notes a few days after a lecture that looked like the ones in Figure 2.1. Notes like these are difficult to understand partly because they are lacking in structure.

Alternatively, the notes represented in Figure 2.2 have better structure and so should be much easier to understand. The main purpose of structure is to make clear which

FIGURE 2.2 Well-structured lecture notes.

Transition Metal Chemistry

Filling d orbitals, called d-block elements.

Electron configurations

Elements - 4s orbitals lower in energy

Fe [Ar] 4s² 3d⁶

Cr [Ar] 4s¹ 3d⁵ - [illegible]

Compounds - 3d lower in energy

Cr(0) [Ar] 3d⁶ eg [Cr(CO)₆]

Cr(III) [Ar] 3d³ eg [Cr(H₂O)₆]³⁺

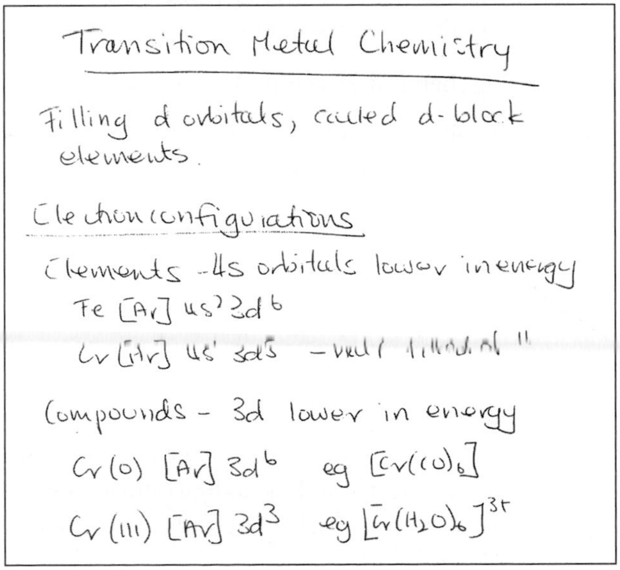

the important information is. It is crucial to note that identifying important information is much easier to do during the lecture because the lecturer will use their tone of voice, pace, and many other devices to provide emphasis to the material, thus giving you an indication of what aspects of the information are most important. After the lecture, however, it is very difficult to recall this kind of detail.

To give your notes structure, you should:

- use headings to order information;
- give each point a new line;
- highlight examples and illustrations in an appropriate fashion;
- use diagrams to summarize information;
- make clear when sections of your notes are digressions from the main points.

Use your own words

One of the reasons why it is important to try to use your own words, when making notes in lectures, is because it will help you (or perhaps force you) to understand the content of the lecture better. If you are trying to put information in your own words, then as you hear it you will need to process it in order to have your own words to put it into. It also helps you to make notes, as opposed to merely take notes (as explained above). You will have probably experienced occasions in lectures when you have been able to write down what the lecturer is saying without actually thinking about what is being said, this illustrates how passive taking notes can actually be. How much you try to put information into your own words will depend to a certain extent on your priority; to understand the information or collect information; however, whichever your priority is, putting information in your own words will help you understand it better. This doesn't mean that you have to put absolutely everything in your own words; for example, there are two particular occasions when it is important to record the precise wording:

- when the lecturer is using chemical or technical terminology;
- when you are recording a quotation that the lecturer is referring to—in which case make this clear in your notes by using quotation marks;
- when you don't understand what the lecturer's words mean—in which case make this clear in your notes by adding a question mark in the margin (for example) as a reminder to follow-up the point later.

Know how much to write down

There are two potential problems related to how much information you write down; writing down too much information or not writing down enough. How much you write down will again depend on your priority (understanding information or collecting information), but it is also influenced by your own level of confidence. Under-confidence tends to lead to writing down too much, whereas over-confidence tends to lead to not writing down enough, neither of which, obviously, is ideal. Amongst first-year undergraduates probably the more common tendency is to write down too much information, which may not be a problem beyond giving you writer's cramp! Whichever your tendency, the following suggestions will help:

- remember that lots of notes don't necessarily equal good notes;
- look and listen for the important points, these are often the structural parts;
- use keywords to represent points or ideas concisely;
- add brief details of any examples or evidence that support a point.

Use abbreviations

Using abbreviations can be a real time saver (as you will know from text messaging). Use standard abbreviations, subject-specific abbreviations and your own abbreviations for common words. The important thing is to be consistent and to ensure that your notes are still comprehensible; don't use so many abbreviations that your notes turn into a shorthand transcript—these can be very difficult to decipher when your memory of the lecture has faded.

Use space

It can be tempting, in an effort to save paper, to try to cram as much information onto a page as possible, but this will create difficulties for you both during and after the lecture. Notes that are densely packed are difficult to review and difficult to make additions to at a later stage. It is therefore helpful to use space in your lecture notes to make them easier to review and easier to supplement with additional material. To create space make sure you put each point on a new line (this also helps represent structure) and leave gaps for additions or corrections, especially if you think you may have missed or don't understand something.

Use colour and image

We have already identified in the section on *Use structure in your notes* that it is important to highlight important points by using structure, but this can be further enhanced by using colour and image too, as shown in Figure 2.3. It can be useful to highlight in colour key points and to use images or diagrams as a quick way of summarizing a concept or idea. Sometimes such images or diagrams will be used by the lecturer, in which case, if it is useful, copy it down (although see the points below on *Use slides and handouts* effectively. On other occasions you will think of ways to represent visually something that is only communicated verbally, this can also be a very helpful thing to do, but make sure you record sufficient information to be able to understand the concept or idea at a later stage.

FIGURE 2.3 Using colour and image in notes.

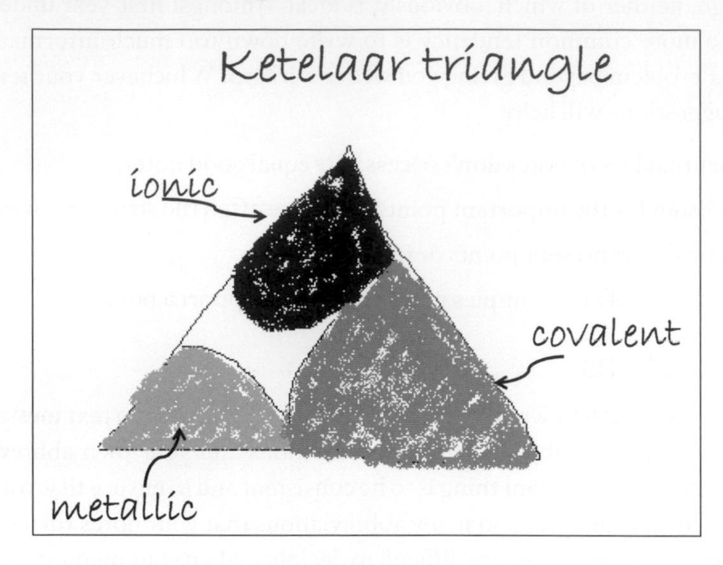

Use slides and handouts effectively

Slides and handouts are a very valuable source of information. Not all lectures will provide handouts or copies of their slides. Assuming that copies of the slides or a handout are available to you (either before or after the lecture) then the principles of using them effectively are similar.

- Put your name on the handout—everyone's will be identical (initially at least), and you will want to know which one is yours.
- Put a date and lecture title on the handout (if there isn't one on it already) so you know which set of lecture notes it belongs to later.
- When making notes, think about what is contained in the handouts or slides and don't copy down in your notes things that are already in the handouts. This will enable you to listen more actively and so focus on understanding.
- Make notes on the handouts themselves—highlight important points, add comments, write down any questions you have.
- Many lecturers will provide gapped handouts or slides where you need to fill in the gaps with important information.
- Don't fall into the trap of thinking that because you've got complete handout or set of slides that you don't need to listen much—listening actively during the lecture and making appropriate notes of clarification in your own words on your handouts will save you a lot of time later and also make it easier to understand the material.
- File your handouts in an organized fashion (which brings us neatly onto the penultimate point in this section).

Organize your notes

It is important that your organize your notes effectively, as this will make it much easier for you when you follow up the lecture at a later stage or use the notes for background reading for an essay, or when revising for exams. The simplest, and in our experience the most effective, method is to use A4 paper for taking notes and to store your notes in a ring binder, along with the relevant handouts. To make ordering your notes easier, begin notes for each lecture on a new piece of paper and give them a clear heading of the lecture title, date and name of the lecturer. Also add page numbers so you can order the pages easily. To store your notes and handouts in a ring binder you will need to hole-punch them—do this before they get lost!

Compare your notes with other people's

Finally in this section, compare your notes with other people's. Remember, we said earlier that the ability to make good notes is a skill that develops with practice; it is also a skill that develops as you see how other people do it. Comparing notes can be a helpful exercise because it can help you to:

- identify and fill in any gaps that you might have in your notes;
- discuss the content of the lecture and clarify your understanding;
- identify how the lecture relates to the rest of the module.

2.4.3 **Know the common problems and how to address them**

We have addressed the ways to help you to make appropriate notes, but there will always be times when, for a variety of reasons, concentration dips and you begin to get left behind in lecture. The trick here is not to despair and give up!

Failing concentration

You are much less likely to find your concentration straying when you use an active approach to note making. Putting points into your own words, using space, colour, and image, will all help make note making a busy but interesting activity. If you do miss some points because your attention strays, then just leave a space in your notes and check it out with the lecturer or another student later. Many lecturers will build in a break or activity into their lectures to allow you time to refocus and regain your concentration.

Being left behind

You may find that information is being delivered too fast for you to write down. If points pass you by, then leave a space and compare your notes with another student's. Doing some background reading for the lecture will help you to keep up as the information won't be entirely unfamiliar to you. Sometimes you can get lost because you don't understand the material that is being delivered. This may be the case for the occasional point or even for a large section of the lecture. Rather than giving up on the lecture, write a series of questions that you can try to follow-up later.

Try this: Practice taking notes outside a lecture environment

Choose an informative television report, such as a news bulletin or documentary. Using the techniques discussed above, try to take concise and meaningful lecture notes over a 15-minute period of the report. Use your notes to produce a fuller report. If you carry this out with a friend you can compare notes afterwards and share them to produce an even better report.

2.5 Follow-up after lectures

One of the things that we said was different between lectures and lessons was that, at school or college, lessons usually cover everything a student needs to know about their syllabus, but at undergraduate level lectures don't tell you everything you need to know and so need to be supplemented with additional reading. It is vitally important that you are aware of this because when you are revising for exams, if you only revise the lecture material your knowledge of a subject will be inadequate. We will deal with this more in Chapter 14, *Getting the most out of revision.*

2.5.1 Know what kind of follow-up is required

In terms of the information covered, lectures are commonly used to either: offer a broad overview of a subject (in which case your job is to fill it the detail); or deliver detailed information on a subject (in which case your job is to fill in the background). There are therefore two issues to consider when identifying what sort of follow-up a lecture requires; does the subject require more depth (because the lecture gave an overview) or does the subject more breadth (because the lecture gave detail)?

2.5.2 Ask questions

Don't be afraid to ask a lecturer for clarification either during the lecture or afterwards. This can take a bit of courage, especially putting your hand up and asking a question in a room full of people, but the chances are, it is very unlikely that you are the only person in the room thinking that particular question. Also, lecturers will often welcome such questions as they provide a bit of stimulus for them and they indicate that people in the room are paying attention. Additionally, use tutorials to clarify or discuss material from the lectures.

2.5.3 **Review your notes**

Review your notes as soon as possible after a lecture, and make the most of your review by:

- highlighting points that seem particularly important or central;
- adding any details that you can remember from the lecture;
- showing links between points;
- correcting any mistakes;
- adding questions to highlight areas you don't understand or need further information on.

* Chapter summary

Making the most of lectures is an important skill for undergraduates. There is much you can do to improve your ability to listen and make notes, this includes things you can do before, during and after lectures. The things you can do include:

- be prepared—get to know the course structure and do pre-lecture reading;
- be organized—have a system for storing notes and take a selection of pens and pencils to the lecture;
- listen for structure—watch out for signposts that help you follow the lecture;
- be brief—try using key words and phrases as much as possible so the emphasis is on listening and understanding;
- make note making an active process—summarize in your own words, make good use of space, colour, symbols and images;
- leave gaps—if you miss a point, don't get further behind by panicking about it, just leave a space and check it out with another student after the lecture;
- actively review your notes after the lecture, making additions and corrections.

Chapter 3

Making the most of tutorials and group work

➲ Introduction

Tutorials and group work form key elements of teaching within most degree programmes and can be employed in a variety of formats, as we will see in this chapter. Irrespective of the specific format of a session, the most important things you can do are to **prepare** and to **participate**. Without preparation and if you are not willing to participate, you will derive little benefit from a tutorial or group work session.

Tutorials can be quite a daunting prospect as many students are unsure of their knowledge base and ability to discuss ideas, particularly in front of a member of staff who may be a world expert on the subject. But you should remember that your fellow students will be in the same position and your tutor will want to encourage you to participate.

Tutorials and group work are useful in developing a range of important skills that will be of value to you in your studies and also in your career development. These skills include the development of:

- subject awareness – increasing your knowledge of the subject;
- critical thinking – listening to different ideas as they are put forward, thinking about them and evaluating them in terms of the strengths and weaknesses of the arguments;
- communication skills – presenting your own ideas in a way that can be understood by the rest of the group, and learning the essential ability to be able to refine the views expressed by other people, pick out the key points, or even counter them, without causing offence.

We will begin by addressing tutorials.

3.1 Tutorials

The basic format for a tutorial is a relatively informal group where a tutor meets with a small number of students to discuss specific aspects of the course or module being studied.

3.1.1 Different types of tutorial

There are two main types of tutorial.

1. Academic tutorials that are normally linked to the modules you may be taking at the time. These sorts of tutorials may very occasionally be run on a one-to-one basis but will more commonly involve five or six students or even up to 20 or so. Academic tutorials may involve a range of activities such as discussions about a specific topic, short presentations given by members of the group followed by a discussion, or the consideration of the answers to previously circulated questions or problem sheets.
2. Tutorials held with a personal tutor, or equivalent, to discuss your progress on the course. These will often be on a one-to-one basis and may include discussion of career issues and personal development planning, as well as aspects of the non-academic areas of university life, such as accommodation.

As well as tutorials you are very likely to be required to participate in various forms of group work. Many of the principles, particularly in terms of engagement, are the same for the two types of session, but we will also consider group work separately later in the chapter (Section 3.2).

3.1.2 Preparing for tutorials

Preparation for tutorials is essential. It may well be that you have been given a specific topic to research, a worksheet to go through, or some sample questions to answer. On the other hand, you may simply have been told what the theme of the tutorial will be. Either way, you need to be prepared, so that you can benefit from the discussions and participate in a meaningful way.

Preparing for academic tutorials

Academic tutorials offer you the opportunity of discussing your subject with a chemist, who is very knowledgeable, as well as with your fellow students, in a fairly informal setting. If you are going to be able to engage in that discussion and to learn from it, then careful preparation is essential. It is likely that you will be given one or more specific topics or a worksheet of problems that will form the starting point for the tutorial discussion. As with the preparation for writing an essay (see Chapter 7, *Writing essays and assignments*), or giving an oral presentation (see Chapter 11, *Preparing scientific presentations*), you need to read up on the topic and make succinct notes that you can use as prompts in the discussion.

If the tutorial topic is based on material that you have been covering in your lectures, then the obvious place for starting your reading will be your lecture notes, before moving on to read any specific references you may have been given. Have a look at the sections on *Using textbooks*, *Reading a research paper*, and *Note-making strategies* that are covered in Chapter 5, *Working with different information sources*. As a guide, the most important things to ensure are that:

- you are keeping focused on the question asked;
- you are making brief notes that summarize the key points and that you will be able to use as prompts in the discussion;
- you are keeping a record of the sources of the information (see the section on *Citations and references* in Chapter 5) so that you can cite the paper or text if you are challenged regarding the reliability of the information you are presenting, or if you want to find the information again to get more detail.

A short time before the tutorial read through your notes to make sure that you understand the points and to refresh your memory.

Preparing for personal tutorials

Personal tutorials may cover a range of topics. In relation to the themes of this book the most important aspects are likely to be topics such as your academic progress, module choices, personal development planning, and thoughts on careers. Again, if you are going to benefit from the opportunity then you must prepare in advance and then be ready to discuss these issues openly.

For example, in order to be able to discuss proposed module choices for a forthcoming semester or academic year you need to have read about the range of choices that are open to you: this will include checking whether taking the modules require you to have taken specific pre-requisite modules in the current year of study. You also need to have thought carefully about where your strengths and interests lie and where you think studying those modules might take you in career terms. In the context of your development, you should also be thinking about whether you would benefit from further training to develop your skills: for example, if the modules are assessed by poster presentations and you are not sure how to go about preparing posters, then you should read Chapter 13, *Creating academic posters*, in this book and also consider seeking guidance from your tutors or perhaps from the university's learning support centre, or equivalent.

3.1.3 **During tutorials**

You have come to the tutorial, having done your preparation. The next important element is to make sure that you participate. Participation, of course, means making active contributions to the discussions, but it also means listening carefully, thinking about what is being said, and relating that to what you have already learnt.

Contributing to academic tutorials

The first and most important thing to remember about tutorials is that they are not mini-lectures: your tutor will expect to have a role in facilitating and guiding the discussion but will not be expecting to be talking for much of the time. Indeed, some tutors are experts at sitting in silence waiting for the students in the group to start talking! There will, therefore, be an expectation that you will be an active participant – that you will attend the tutorial having prepared beforehand and be prepared to discuss your ideas, as well as to listen to the points made by your peers and the tutor.

You will probably find that this is all quite a daunting prospect at first and there is a strong temptation to sit looking at the floor, hoping not to be noticed! While this is an understandable response, it is not a very productive approach. Remember that your tutors are there to help you understand the subject and will want to help you develop your skills in communicating ideas, however, they can only do that if you participate. Don't be afraid of expressing ignorance if there is something that you don't understand: the only thing that is likely to be irritating to the staff is if it is apparent that you have not bothered to do any reading or preparation beforehand.

Depending on the structure of the tutorial, there may be some short presentations about topics that you and your peers have been asked to prepare in advance, followed by discussions. Alternatively, there may be specific questions or problems that you have been asked to work through, so the whole tutorial takes the form of a facilitated discussion. Irrespective of the format, there are key ways of engaging with the process.

- Listen carefully to any ideas being presented by the tutor or other students and note down the key points of the arguments.
- Be prepared to say whether you agree or disagree with the ideas, presenting evidence to support your views. It is no good just saying you agree or disagree with an idea: you also have to be able to justify that view.
- If there are things you don't understand, then ask for further details or explanation: never leave the tutorial not understanding things because you were afraid to ask questions.
- At the end of the tutorial, make sure you can summarize the key points for yourself.
- You do need to be sensitive, though: if you are a confident speaker, don't try to dominate the discussion but let other people have their say as well since that is an important way of learning.

Try this: Coping in tutorials

Decide how you would deal with the following situation:

You have turned up to your next scheduled tutorial. There are four students and a tutor. You have prepared well by going through all the problems and questions on the sheet provided beforehand. You are looking forward to getting some feedback on your efforts. The tutor runs the session by asking for answers to questions and you all go through the sheet. One of your fellow students is very vocal and overbearing and quickly jumps in and answers every question. Your tutor just seems grateful that someone is contributing and does nothing to stop him.

Contributing to personal tutorials

With personal tutorials, as with academic tutorials, you can only expect to benefit if you engage actively in the process and, in these cases, are prepared to be open in terms of your views and aspirations, as well as being realistic in terms of your abilities. This means that you may need to open up discussion topics and explain your background to them, rather than waiting for your tutor to raise them: despite their best intentions, they may not be able to guess what is on your mind or worrying you at the moment. You should also recognize that your tutor may not have all the answers. Even if they don't have the answers, however, they should be in a position to be able to advise you on where to seek further guidance.

It is important to remember that the personal tutor is there to try to help you. Often students are anxious that if they raise issues of concern with their personal tutor, it will affect how the department views them. This should not be the case under any circumstance. Although the personal tutor is not bound by a confidentiality clause as are, for example GPs or counsellors, they have a duty to be discreet about things they are told and not to be judgemental.

Try this: Getting help with academic problems

How would you deal with the following situation?

You are having serious problems with the course and think you may have chosen the wrong modules, or even the wrong degree? You know that you should discuss this with your personal supervisor but you find her very difficult to talk to and keep putting off making an appointment.

3.2 Group work

The process of collaborative or group working may often be very productive. Indeed, there is a wealth of evidence that material learnt through group work is retained better than that learnt through formal teaching methods. Group work is often very useful in developing critical skills because as a member of a group you need to evaluate the contributions that other members of the group have made, as well as reflecting on your own contributions and discussing their validity with the rest of the group. In the chemical sciences, group work is very common as a part of laboratory classes where you typically may have to work in pairs or threes in order to be able to complete the experiment. In the case of group work based around discussion questions, many of the points relating to tutorials may also be applied. In particular, you again need to be thorough in your preparation and be prepared to engage actively in discussions.

One of the additional key features, though, relates to organization and the dynamics of interaction of the group. Unlike the tutorial, where your discussions are facilitated by a tutor, group work exercises will normally require you to work with your group on a set task, either in a classroom setting, for a relatively short time, or as an extended exercise which you undertake at times and locations of the group's choosing. The dynamics of how people interact and contribute to the group are a very important element contributing to the overall success of the group. Under these conditions, how you behave in the group is also clearly important. Some key points to remember here are:

- keep an open mind and be prepared to listen to, and consider the ideas put forward by others in the group;
- avoid being judgemental, particularly, for example being dismissive of someone else's ideas;
- be careful of your body language, for example, folding your arms and staring out of the window does not convey a positive message.

Let's now consider each of these types of group activity.

3.2.1 Short-term group work

Short classroom exercises are often employed as a way of breaking up lectures or as a means of checking that students have understood a specific concept. You will probably be asked to talk to the people around you to discuss a specific topic or answer some questions. Many of the issues of group dynamics and team playing that are discussed in the next section are not applicable here. For the most part, the most important skill is remaining focused on the topic and using the time constructively: it is very easy for discussions to degenerate into a review of the latest gossip and, whilst that might be of great interest, it will not help get the task completed! It is also often very easy to spend a long time debating the first one or two aspects of the question and then find you have run out of time before completing the exercise, particularly if you only have a few minutes for your discussion.

There are several ways of providing some structure to your discussions and making sure that you get some answers whilst also getting everyone involved as far as possible.

- If you have been given several questions to answer, it is very helpful to agree quickly at the start how long you are going to spend on each question, and make sure that you adhere to that timescale, noting down your agreed answers as you go.
- A useful way of getting ideas quickly and of engaging everyone in the discussion is to go round the group in turn with everyone putting in an idea or point of information related to the question.
- An alternative is to brainstorm: people in the group mention ideas in any order as they think of them. This can be very productive if some creative thinking is needed, because the lack of formality encourages more open thinking. It can also generate numerous ideas very quickly. However, it may also lead to some people feeling excluded because, for example, they feel apprehensive about the merit of their ideas or because they are too shy to contribute. In such situations, try to be aware of the other members of the

group and encourage non-contributors to participate: after all they might have the best ideas of everyone! As with the guidance for behaving in tutorials, the key points here are to listen to what others have to say, to ensure your body language is appropriate and open and not to be dismissive even if you disagree with what the other person is saying.

3.2.2 Longer-term group work

If you have been allocated a task that will require you to work in a group over a period of days or weeks in order to complete the task, then the dynamics of the group become a very important factor contributing to the success, or otherwise, of the exercise. Many students are dubious about the benefits of group working, particularly if the exercise is worth a significant amount of marks. However, the skills involved in such exercises are very important in career development because team-working skills are among those most highly valued by employers.

Undertaking a specific project as a group can be broken down into a series of key stages, as shown in Figure 3.1. We will consider each stage in turn.

Get off to a good start

It may be that you have been allowed to select who you work with, and have chosen to work with a group of friends. On the other hand, you may have been put into groups by your lecturer and you may or may not know the people you are having to work with (and, if you do know them, you may not like them!) You might not feel happy if you find yourself grouped with people you don't know, or don't like, but it can be very good experience for your working life, since you may well have to team up with strangers to undertake a task in your employment, or you may have to work with people you don't necessarily like. In either case, the work still has to be done.

Time spent in ensuring that the group gets off to a good start should pay dividends as the project develops. Setting up formal structures for the group – like having a group leader, agreeing task allocations, or having meeting agendas – might seem rather over the top for some projects, but it can all be useful experience. Additionally, having formal aspects to the project, which are agreed in advance, help in avoid problems later on.

Get to know each other

This first stage is clearly most important if you are going to be working with people you don't know. Even in a group where *you* know everyone, that might not be the case for everyone. So, it is very helpful to get together for your first meeting, away from the work environment: perhaps in a coffee shop or pub or just sitting on the grass if the weather is good! Use that meeting to introduce yourselves and ensure that everyone feels part of the group and can, at least, begin to feel confident about working together. If you need to do a round of introductions, get each person to say something about themselves as a 30-second biography or say what they think might be particularly interesting about the project.

FIGURE 3.1 Stages in the process of group work for a specific project.

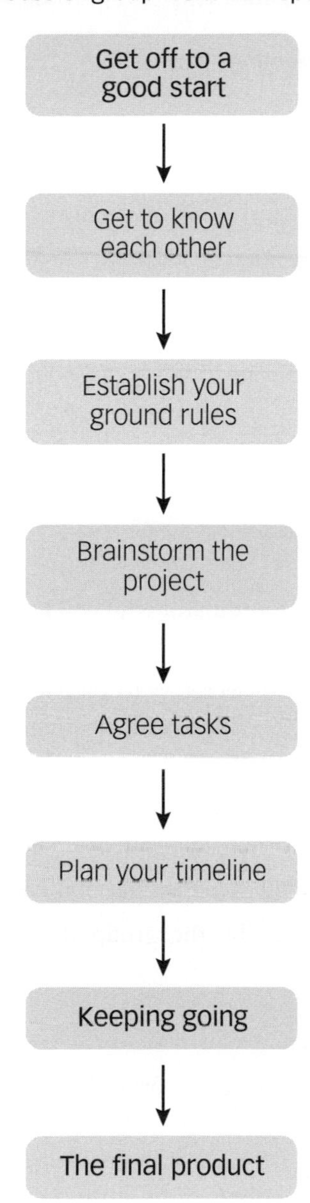

Establish your ground rules

Don't be misled into thinking that, just because you are working with friends, you can all just get on with the task. Whether you are with friends or strangers, it is essential that you agree the ground rules for running the project, otherwise things can come unstuck very quickly. The extent of these rules will depend on how long the project will be running for. If it is several weeks, however, you might like to consider the following questions/points that could be set out in a generic agenda for the first meeting (as shown in Figure 3.2).

FIGURE 3.2 Proposed plan for the group's first meeting.

Med Chem Project Group Meeting Plan
14:30 Tuesday 2 May
Union coffee bar

Introductions
Group organization
Do we need a group leader?
Do we need a secretary?
Do we need...
Group operation
Brainstorm
Division of project tasks
Key progress dates
Next meeting

- Do we need a project leader/coordinator who will keep track of progress or who can chair group meetings?
- Would someone take the role of secretary to keep a record of points agreed at meetings (as shown in Figure 3.3), and make sure meetings stick to time (it doesn't have to be the same person each time).
- Do we need formal agendas for future meetings? If you think setting agendas makes the whole process seem too formal, then you could have meeting plans.
- What time commitment should each member of the team be making to the project?
- What action should be taken by the group if one or more members are not contributing?

Brainstorm the project

You have been given the project title and a brief for what is expected of you. No doubt different members of the group have different ideas about what the approach should be, what needs to be done, and how. Use a brainstorming session to let everyone have their say. It is important that the discussions at this stage are free ranging, so that the full breadth of the topic can be covered and also that everyone feels free to participate. So, at this stage, all contributions should be treated equally and there should not be any criticism of the ideas that are offered: you will have time later on to focus in on specific aspects.

Note: it is important that someone records the ideas, otherwise all that brain power will be going to waste! However, you need to make sure that the person recording ideas has the chance to contribute ideas as well and is not just there as a note taker. Also, set a time limit for the discussion, otherwise you could be there a long time!

FIGURE 3.3 Example of first section of notes from the group's first meeting.

Med Chem Project Group 1st Meeting Report
14:30 Tuesday 2 May
Union coffee bar

1. Present
Alex, Komal, Sarah, Lee, Rathan

2. Organization
Komal will act as group leader, Sarah will take notes

3. Operation
Project task is to present a critical review of the various stages in the development of a new pharmaceutical product.
6 weeks working time.
Group to meet every week to update

4. Brainstorm
What sort of pharmaceutical – does it matter?
How identified initially
Do we have to think about legislation etc?....

A useful approach to brainstorming is to give everyone two or three minutes to think about the problem and then go round the room so each person puts their ideas in. These ideas can be recorded as bullet points on a flip chart that everyone can see. When the first round of ideas is up, see if there are links between them so that you can group them together. From that grouping, you can then go round the room again to build on these with further ideas: but remember your time limit!

Agree tasks

Once you have the outline of how to approach the topic, you need to break this down into a series of tasks, so that you can plan the way the work will be done and who will do it (as shown in Figure 3.4).

One of the obvious features of group projects is the division of labour so that there is cooperation; if you have everyone trying to do everything, there is no real benefit from being in a group. Again, it is vital that, right from the outset, the group agrees who is going to do what and that a common record is kept of the division of labour (Figure 3.5). So, having drawn up a list of tasks from your brainstorming session, the group needs to agree how they are divided up so as to be reasonably equitable. At this stage it is really helpful to take account of different people's particular skills: if you have a successor to David Attenborough as a presenter, then it makes sense for that person to be taking a major part in any presentations.

FIGURE 3.4 Identification of the tasks involved in preparing the presentation.

Tasks for drug design project

Research topics

1. Identifying lead compound – how?
2. What would initial expts be? Identify active ingredient?
3. Analogues? Synthesis
4. Toxicology, trials, registration
5. Is there a market? How do we market? Economics?

Identify the key points for the presentation
Prepare the presentation and write the text
Prepare handouts for the audience
Rehearse and review the presentation

Plan your timeline

As well as dividing the project up into different elements that individuals will be responsible for, you also need to divide it up into chunks of time so that you have specific milestones that you can identify in order to record progress. It may be helpful to link this in with the agreed division of labour, so that each person has a target to work towards and take responsibility for (Figure 3.5). Also, build in some slack in the timing, so that you allow some extra time at the end of the project for completion. For example, you could aim to complete the project a week before the deadline, then if there are any problems you do have some leeway.

Keeping going

It is often the case that everyone is pretty enthusiastic about the project to begin with and the first meeting goes well with agreed plans of action and so on. Keeping the momentum

FIGURE 3.5 Key dates for the project and breakdown of project tasks.

Project timeline:

May 9th	Completed background research
May 16th	Agree the bullet points for the presentation
June 6th	Draft presentation complete
June 10th	Rehearse presentation
June 13th	Class presentation!!!

Work division:

May 9th

Komal – background, pharma industry
Alex – lead compound, initial chem. and toxicology
Rathan – what is needed to get to market?
...

going, however, can be much harder. This is especially true when more imminent coursework deadlines focus input elsewhere and the long-term nature of the group project tends to give rise to an element of complacency. This is when it becomes important to monitor – and stick to – your original project plan.

Meet regularly

Keep to your schedule of regular meetings. In particular, don't let a meeting be dropped because people have not done their work (even if one of those people is you!): it is very easy to, and you will quickly find you are on a slippery slope and no progress is being made. If nothing else, therefore, the meetings act as a reminder that the work needs to be done and that the group needs to keep on schedule otherwise the project will be compromised.

Keep a check on progress

At each meeting, it is important to review exactly where the project is and whether you are on track. Refer back to your timeline (Figure 3.5) and check off your progress against the target dates.

This checking, however, is not just a tick-box approach, blindly accepting that the work has been done. If the group is aiming for high quality in its performance, then there also needs to be sharing of information and evaluation of the quality of the work each person has produced. If your tutors gave you a marking scheme or equivalent, then the members of the group need to evaluate each section of the work that has been produced against the marking scheme. Though it may seem like it, this is not just a chore: it is important for your own learning that you have an understanding of the whole project, not just the sections you are responsible for. It is also a very valuable learning exercise in developing the skills of critical evaluation, reflection, and discussion.

People who fall behind

One of the biggest, and most common, problems with group work is dealing with group members who don't keep up with the work, or fail to engage with the group. There are various strategies the group can adopt to try to resolve the situation.

- Talk to the person. This is the first and most vital approach. If the person comes to the meeting, then you can discuss the situation. If he or she has stopped attending the meetings, then you need to try to re-establish contact as soon as possible.
- Find out what the problem is. It may be that the person has been ill, or struggling with something else and can catch up, particularly if given a bit of time or some help on a one-off. If there is no real excuse, then the group needs to refer back to the ground rules and make it clear that everyone needs to fulfil their agreed contributions, otherwise the whole project may fail.
- Persistent non-cooperation. If one member of the group will not cooperate, even after you have discussed the position, or if that person refuses to respond, then you need to take more explicit action. Probably the best option here is for you to contact the tutor and explain the situation and ask for assistance in resolving the problem,

though this may be a move of last resort as it indicates that the group can no longer sort out its own problems.

- It may be that your tutor has built in some peer assessment into the group project. Peer assessment allows each student to provide a mark for each other student in their group that reflects the relative effort that each has put in to the task. This mark counts toward the final mark for the project. This may be the best way of ensuring everyone pulls their weight and, if all else fails, at least it gives you the opportunity to penalize the shirkers!

The final product

The deadline is approaching and you now need to bring everything together ready for submission. If you have let things slip, then the quality of the final product may be lacking. If, however, you have kept to your timeline, this phase of the project should allow you to produce a really polished piece of work.

- Bring everything together: one of the difficulties of group work is that everyone has their own style of writing or expressing themselves but the final piece of work needs to appear as a consistent piece and not as three or four completely separate elements that have been lumped together. This may require you to edit the sections so that the approach or style of writing is the same and also so that there are sensible links between the separate sections. Editing by committee is often a lengthy and unsatisfactory process. In most cases, it is preferable for one person to be responsible for bringing the sections together and editing them to create a single 'voice': think back to the process you went through to agree the tasks so you make sure that you play to the strengths of each member of the team.
- Rehearse your presentation (see the section on *Practice* in Chapter 11, *Preparing scientific presentations*): if your final product is a presentation, then it is essential that you rehearse. In particular make sure that:
  - the presentation keeps to time;
  - the illustrations all show up appropriately. This is especially important if you are using animations or, for example, showing video clips that are embedded in the talk;
  - if you have more than one speaker, make sure that the links between them are smooth and everyone knows who is speaking in turn.
- Check against the brief: before you submit the work, run through the instructions you were given at the outset again and make sure that you have addressed all the criteria. It is important that, as far as possible, everyone agrees that they are happy with the final product and are prepared to sign it off.

When it is all over don't forget to make use of the feedback (see Chapter 16, *Using feedback*) and think about where any problems were and how you might do it better next time.

✱ Chapter summary

In this chapter, we have looked at ways of working for tutorials and group work. The key message for all these activities is that you need to prepare, to participate, and to listen to others.

When taking part in academic tutorials you should:

- be prepared and carry out any work that was set prior to the tutorial;
- read through any lecture notes or reference material in order to make sure that you know the subject well beforehand;
- think about what the objective of the tutorial is;
- make sure you listen to what others have to say and be prepared to take their ideas on board as well as argue against them if necessary;
- take notes, so that you have a record of the discussions.

For group work you must:

- decide whether you need to agree a leader;
- be prepared to work with other people, even if you don't like them;
- agree the tasks to be done, who will do them (work to each person's strengths) and by when;
- keep a track of how the project is developing by assigning a note taker;
- meet regularly and check progress against your timeline.

Chapter 4

Making the most of practical work

➲ Introduction

Practical work is probably the defining feature of an education in the sciences. The knowledge base in all areas of chemistry is underpinned by careful experimentation and observation. The abilities to design experiments, to undertake them methodically, to observe and record the outcomes accurately, and to analyse and interpret the findings are vital to the development of a scientist. Students in the chemical sciences may spend many hours carrying out practical work and it is therefore essential that you are able to make the most of the opportunities presented to you. During your studies you are likely to undertake a range of practical work, including practical laboratory work and independent project work. The aims of these types of work are to develop your skills as a scientist and also to deepen your knowledge and understanding of the subject. This chapter will explain the aims of practical work and help you to approach it in the most effective way. We will focus on laboratory-based practical work but recognize that many practical activities may not take place in a laboratory. Much of the material is also directly relevant to project work as well. To help create some context, the text uses as most of its examples an undergraduate investigation of the hardness of tap water by analysing the concentrations of calcium and magnesium ions by a complexometric titration. The chapter looks at preparing for a laboratory class and carrying out the experimental work, as indicated in Figure 4.1. Writing up practical work is covered in Chapter 8.

4.1 The aims of practical work

There are many reasons why chemical science programmes include long hours of work in the laboratory. First and foremost, chemistry is an experimental science; practical work is what chemists do. It is too easy to see chemical science as what you do in lectures rather than what you do in the laboratory. So practical work gives you a taste of what it is like to be a chemist. In terms of specific skills, the most obvious aim is to enable you to develop

FIGURE 4.1 Structure of the chapter.

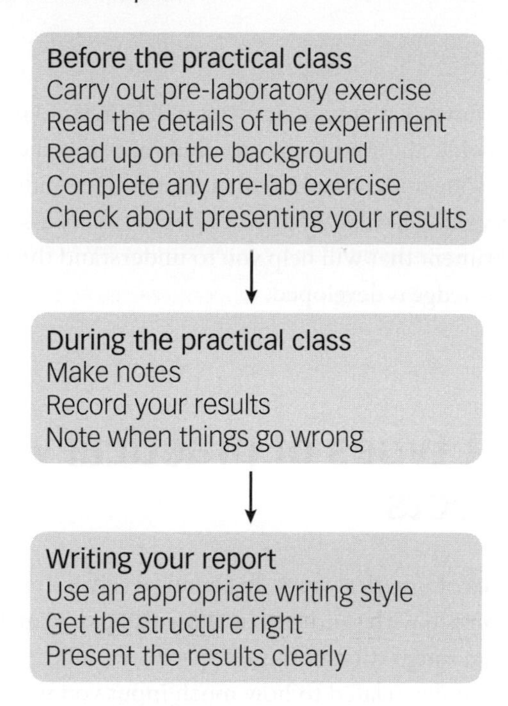

the manipulative and technical skills that are necessary for anyone working in a laboratory in a professional role. But there are many other reasons for doing laboratory work.

> **Try this: The aims of practical work**
>
> List as many reasons as you can for including laboratory work in your course.

Here is a list of skills that can be developed though practical work in the chemical sciences. This list is by no means exhaustive:

- manipulative skills;
- use of instrumentation;
- confidence in a laboratory setting;
- make and record observations;
- team working;
- communication;
- problem solving;
- time management;
- independence;

- interpretation of data;
- numeracy;
- analytical thinking.

In addition, the experiments that you carry out will be related to your lecture courses and will, therefore, provide an opportunity to illuminate the chemistry you have been studying and enhance your understanding of concepts. Depending on the style of the activities that you carry out you may also be developing skills associated with planning and designing an experiment that will help you to understand the nature of science and how new scientific knowledge is developed.

4.2 Different types of practical work and projects

There are different types of practical work. In chemical science experiments may be categorized as being concerned with synthesis (making things) or analysis/data acquisition (measuring things) or an integration of these approaches. The experiments themselves may also have different styles related to how much input you will have into the experimental design.

Laboratory activities found in the chemical sciences can be useful categorized into four types: expository, inquiry, and problem based (Domin, 1999). For an expository experiment, you would follow detailed instructions in a laboratory manual or script. Both you and the tutor would know what the outcome of the experiment was going to be, for example, a sample of a certain organic compound. These types of experiments are very widely used, especially in the early stages of a course as they are very good at developing manipulative skills and building confidence.

In the guided type of laboratory no detailed laboratory manual is used and the tutor provides minimal guidance but guides the student towards the outcome, which would be known to the tutor but not to the student.

In an inquiry-based laboratory activity the final result or outcome is not already known to you or your tutor. For example, you might be trying to synthesize a molecule that had not been synthesized before. The student would be expected to generate their own practical procedure, perhaps based upon previous work in the literature, rather than just following detailed instructions. This type of activity is more commonly called a project and is particularly common in the final year.

In problem-based laboratories, students are provided with a real-life problem with little data. Students have to identify gaps in their knowledge and ultimately design and carry out an experiment to solve the problem. Problem-based laboratories are closely related to inquiry-based laboratories in that students have to develop their own experimental method.

Try this: What types of practical?

Go through your laboratory manual and identify each experiment as being expository, inquiry, guided, or problem based. Select one of the experiments and draw up a list of the skills that you think you will develop during it.

Regardless of what type of practical work you are presented with it is important that you are prepared to get the most out of the experience that you can.

4.3 Preparing for practicals

Whether you are going to go in to the laboratory and follow detailed instructions or whether you are going in with a method that you have devised for yourself, it is important to be well prepared. Far too often students enter the laboratory without having a clear idea of what the aim of the session is or what they are expected to do. Good preparation beforehand is therefore a key stage to success in all types of practical work.

Much of this section will seem to be simple common sense but if you want to get the most out of the practical work you are going to do, it is essential that you prepare properly. This is something that many of us are not very good at: it is very easy just to arrive at a class and hope everything will be made clear as you go along. However, not only will effective preparation help you learn more and get better marks, but it can often pay off in terms of saving you time and effort when you are actually doing the work. There are few things more frustrating than having spent a long time setting up an experiment only to discover that you have set it up incorrectly or not recorded all the data you need, so you have to start all over again.

In particular, good preparation means knowing:

- where you have to be and when;
- what items you need to bring with you;
- what you are going to be doing and why;
- some background to the investigation;
- how you are expected to present your findings.

For many practical exercises you will be given a schedule beforehand that provides most of this information. Many institutions now routinely use pre-laboratory activities to prepare students for laboratory work. These 'pre-labs' typically require you to read through the manual carefully to identify key information. You may be required to carry out a safety assessment or to do some preliminary calculations. An extract from a possible pre-lab activity for our complexometric titration experiment is shown in Figure 4.2. Whatever the detail of the pre-lab is, you will undoubtedly be better prepared for having carried it out. You will not only learn more whilst in the lab but you will also be

FIGURE 4.2 Extract from a pre-laboratory exercise.

Determination of calcium and magnesium in tap water

1. Calculate the relative molecular mass of the disodium salt of EDTA.
2. What is the colour change you will be looking for in the determination of a) calcium b) calcium plus magnesium.
3. What is the reacting mole ratio in the reaction between $EDTA^{4-}$ and Ca^{2+} and Mg^{2+}.
4. Convert 2.56×10^{-3} mol dm^{-3} of calcium ions to parts per million.

more efficient, probably completing the experiment more quickly, and certainly more safely. Pre-laboratory exercises are usually designed to get you to carry out most of the following steps.

4.3.1 Read the schedule

Reading the schedule carefully will enable you to check the time and location of the class and what items, such as a lab coat, calculator, etc. you need to take to the class. A lined notebook for recording your results and any notes about the experiment is always useful and much better than a borrowed sheet of file paper. It is good practice to get into the habit of writing your experimental notes in a notebook: this will be particularly important for a research project and many scientific employers require you to keep detailed notes so that, if necessary, results can be verified at a later date.

4.3.2 Read the details of the experiment

Reading the details of the experiment will give you a good idea what you will be doing. It may well be that the techniques and equipment will be new to you but familiarizing yourself with the description of the procedures will help when it comes to doing the experiment. Highlight any key sections of the methods, for example, if specific measurements need to be made at specific times: Figure 4.3 shows an example of how you might do this for an experiment. If you are going to be working in a group, getting together to plan how you are going to organize yourselves and divide up the workload will make you more efficient and save time.

FIGURE 4.3 Using highlights to identify key parts of the experimental protocol.

Preparation of a standard 0.01 M EDTA solution

Weigh out accurately to 4 decimal places 0.9–1.0 g of AnalaR $Na_2H_2X.2H_2O$, transfer this to a 250 mL volumetric flask, dissolve it in deionized water, make the solution up to the mark, and mix it thoroughly. Record the weight on your answer sheet.

4.3.3 Read up on the background

Reading up on the background to the work will help you understand the chemistry involved and can put what you are doing in context. It will help to read through your notes from related lectures. If there is a reading list, try to read through at least some of the material beforehand and make notes of key points (see Chapter 5, *Working with different information sources*). There is also the bonus that, if the papers or texts are not available online and need to be borrowed from the library, you won't have to wait because other people have borrowed them first!

4.3.4 Check if there are instructions about presenting your results

Checking how you should present your results is important because it may vary from class to class. Sometimes you will simply be required to fill in tables or draw graphs on the schedule itself and answer some specific questions, other times you will need to write a full report similar to a scientific paper. Again, knowing in advance what is expected of you will help you be more efficient in your work and save you time in the long run.

4.4 During the practical class

If you have done some preparation for the practical class it makes it much more likely that the class itself will go well. There are two important things you need to do during the class to ensure you can write your report afterwards: make notes and record your results.

4.4.1 Make notes

Try to get into the habit of making notes during your practical classes. Many practical classes will begin with a briefing session addressing a variety of aspects of the work, such as safety issues and instructions and demonstrations of the techniques to be used, as well as some background subject information. There will be other times during the practical class when it will be useful to make notes as well, for example, if you need to make changes to the procedures during the course of the experiment.

Before beginning the experiment, check through the sequence of tasks that you have to complete and, especially for timed measurements, make sure you have a clear list of what needs doing and when.

At the end of the class, you may well have another briefing session. Make the most of the information you are given because this could be very important for when you are writing up: make good notes; don't just rely on your memory!

4.4.2 **Record your results**

Recording your results is clearly a critical part of undertaking any piece of practical work. You may find that in experiments involving gathering numerical data you are recording experimental data directly into a spreadsheet or other program. But whether you are storing the information in a spreadsheet or in your laboratory notebook, you still need to draw up tables for the results, making sure you identify what is being measured and what each column of the table represents, and clearly label any graphs that you produce from that data. In a synthetic experiment when you are making a compound, you should record information such as yield and appearance of the product.

During the experiment record your results and observations carefully, clearly identifying numerical values and the units in which they were measured if appropriate. Also, make notes of any changes you made to the protocol, or to the way in which the results were measured. Again, care taken at this stage will save you a lot of time and frustration

FIGURE 4.4 Two very different ways of recording your experimental data.

A Hardness of Tap Water Practical

11/10/10

$Ca^{2+} + Mg^{2+}$

titres 26.10 ml

26.20 ml

26.15 ml

~~Mg~~ Ca^{2+} only

25.70 ml

25.30 ml

25.45 ml

B 1st sample 2nd

~~25.90~~ 25.40 24.50 24.20 24.30

25.80

later on. You are also much less likely to make mistakes. Figure 4.4 illustrates two very different ways of recording the results from a simple practical class measuring the calcium and magnesium concentrations in tap water.

Compare the two sets of records in Figure 4.4 and list the aspects that you think are unclear in the data sets in sample B. If you were presented with these two data sets, would you be able to write about them in a meaningful way?

4.4.3 When things go wrong

What if things go wrong? By their very nature, experiments may go wrong and you find you don't have any data or that it does not make sense. If this happens speak to your tutor as soon as possible, and they will advise you on what went wrong and whether to begin again. In some instances they may provide you with a set of results that you can then use for your report.

4.5 Group and problem-based work

It is very likely that you will find yourself working in pairs or small groups in the laboratory. Group work was discussed in some detail in Chapter 3 so there is no need to cover all aspects once again. However, group work in the laboratory is often one of the more rewarding ways of carrying out practical work, especially if it is used to tackle projects or problem-based activities. Group work in the laboratory is probably the closest approximation to how scientists work in industry,

As with any type of group activity you are unlikely to be able to choose your teams. It is essential that every member of you team contributes effectively if you are going to reach your collective goals. So the approaches we discussed in Chapter 3 are even more important. It is essential that you all:

- get to know each other;
- establish your ground rules;
- decide whether you need to assign a leader or coordinator;
- write down what you all agree to do;
- divide up the work between you, taking into account any skills that team members already have;
- come back to share results;
- agree how to deal with anyone who hasn't contributed or how to fill the gaps in your results.

Group work in expository-style practicals should be fairly straightforward as you know what is expected of you and you can distribute tasks between the group members.

Inquiry or problem-based practicals may prove more challenging. In these approaches it is unlikely that you will have been given any form of experimental protocol. It is unlikely that you have all the information that you require. So, before you even think about the practical work you will have to look up background information, refer to the literature, find out what equipment is available to you, carry out risk assessments, and only then develop your own experimental protocol. You will then carry out the experiment, reflect on the results and repeat, modify, or improve as necessary. If you have a group of people involved in this process it is vital that you communicate effectively together and cooperate in gathering and sharing information, listen to each other when design the experimental work, and use the best set of skills available to you to carry out the procedure. You may find that not all of you are equally involved at each stage. As long as each person's overall contribution is equitable that should not be a problem, and reflects real life.

* Chapter summary

Practical work should be one of the most rewarding and enjoyable aspects of your study of chemical science. The key to getting the most out of the experience is to know what is expected of you and what you can gain from the experience. So in order to maximize learning in the laboratory you should always

- make sure you know where you should be and when;
- take all necessary equipment with you;
- carry our pre-lab activity;
- read through the manual;
- look up anything you are unsure of;
- keep careful notes of results;
- note down how you deviate from protocol;
- ask for help if you need it.

Reference

Domin, DS. 1999. A review of laboratory instruction styles, *Journal of Chemical Education*, **76**, 543–547.

Chapter 5

Working with different information sources

Introduction

In previous generations, students were said to go to university 'to read for a degree', the implication being that most of their working time as students would be spent reading up on the subject they were studying. Even in subjects like chemistry, which have a strong emphasis on practical work, reading is still a very important, if often under-rated part of studying.

What is the purpose of reading? This is an important question because there are several reasons why you will need to read during the course of your degree and the approach that you take will depend on what your goal for the reading is. Some key reasons for reading include the following:

- background reading to find out more about the subject in general;
- following up on lectures, or other teaching activities, in order to help you understand the subject matter you have been taught;
- following up on lectures to develop a deeper and broader appreciation of the subject matter;
- researching topics as part of the preparation for a coursework assignment such as writing an essay or delivering a presentation.

These activities require different approaches to reading because there are very different goals associated with them. For example, the sources that you use and the strategies employed for finding the information will differ, both depending on the goal for your activity and the level you have reached in your course.

In this chapter, we will explore how to get the most from texts, journals, and the internet, and how to evaluate the academic value of a variety of sources. To do this, we will review:

- the types of publication available and how you might use them;
- using internet sources;

- referencing and bibliographies;
- different reading goals and how you approach them;
- note-making strategies.

5.1 Different types of printed publication

There is a wide range of types of publication available for use at different levels of study. Each of these has different characteristics and will probably be used in different ways. In this section we will consider the main forms of publication, some of their key characteristics as study aids, and suggestions as to their usage. Further consideration of the usage of publications and internet sources are given in the sections on note making and reading goals.

A fairly simple way of categorizing the types of publication you might want to use for your studies is as follows:

- textbook;
- specialist topic book/monograph;
- general science journal;
- specialist review article;
- research paper.

Let's consider each of these in turn.

5.1.1 Textbooks

The textbook is probably the form of publication with which you are most familiar from your studies at school or college. Textbooks still form a very important resource for students at university, predominantly in the first and second years of study. Whereas many of your textbooks will have been broad-based (for example, a textbook covering all of chemistry), the books you will be using, even in your first year of study, will be more specialized, focusing on a specific subtopic of chemistry such as inorganic, physical, or organic chemistry. The principle is the same, however: the textbooks offer a broad-based foundation to the study of a subject.

Buying textbooks

Even within the individual topics such as inorganic or organic chemistry, if you go to a bookshop or search online you will find a bewildering array of books available to buy. It is important to be cautious when purchasing books otherwise you can find that you have spent a lot of money very quickly, sometimes on buying books that will not be of great use to you. You have clearly taken the first step in the right direction by buying this book, now you can proceed with caution!

- Most universities publish lists of recommended books for specific modules or courses, so start off by consulting the lists for the modules you are taking.
- Be selective: for example, is it worth buying a book for £40 or £50 for a subject area that you have to take as a single module in your first year, but which you do not intend to continue with later on in your course? If the university library has a number of copies available, you can probably manage without buying one for yourself.
- Having decided you do want to buy a book, you have to choose which one. It may be that your reading list only suggests a single core text, which makes life easy. But there may be a choice of recommended texts, perhaps covering the material to different levels and at fairly different prices. Again, the library is a useful starting point. Have a look at the different texts in the library: you will probably find that they are different in terms of their layout, the way diagrams are used, the level of detail, etc., so you may find it helpful to test-read parts of the book to find which style suits you best. You may also find that they differ in terms of the additional features they offer: for example, is there an associated website with more information or access to self-evaluation tests? Is there an accompanying CD or DVD?
- Wander past the student notice boards of your department or students' union and you will probably see numerous adverts for second-hand books. There may well be a second-hand bookshop. The books should be cheaper than buying brand new (though that is not guaranteed), so you could save money but you need to be cautious. For example, check that it is the current edition of the book that is on sale. If there is an accompanying CD, is it still there? If the book has an associated website, make sure that you can still access it: many sites have pin number access, which is unique to a single user and cannot be transferred if the book is sold on.

Using textbooks

Textbooks commonly provide a broad overview of a topic and can have several uses as study tools. Think about the reading goals that you have and that will guide you in the way to use your textbook. Some possible examples could be:

Goal 1

I need to broaden my knowledge and understanding of a specific topic, for example nuclear magnetic resonance (NMR) spectroscopy.

Action 1

In a general organic (or analytical) chemistry textbook read through the chapter(s) on NMR spectroscopy, checking on any learning outcomes identified in the text. Make abbreviated notes to aid your learning (see Section 5.3 on *Note-making strategies*).

Goal 2

I have just had a lecture on the interpretation of NMR spectra and I don't understand it!

Action 2

Read through your lecture notes and any accompanying handouts. Look up NMR spectroscopy in your organic chemistry textbook and read through the relevant section slowly. Then read it again, making notes about the principles of interpretation and how they are applied. It might help to go to the library and read about it in a couple of different textbooks. You should also ask the lecturer for help, but it is a good idea to try to think it through for yourself first.

Goal 3

I have just been given an assignment in which I have to identify unknown compounds using NMR spectroscopy and mass spectrometry. Where do I begin?

Action 3

Read through any relevant lecture notes and handouts first. Then read the sections in your textbook that deal with NMR and mass spectrometry, making notes of key stages in the interpretation process. Have a go at all of the relevant worked examples and end of chapter exercises, checking your answers as you go. If you are still having problems you may need to refer to more detailed sources such as a specialist topic book.

As with any source of information, textbooks do have limitations, as follows.

- Because most textbooks offer a fairly broad coverage of the subject (e.g. inorganic chemistry) they may well only provide a fairly superficial overview of a topic. This is important because it may be just what you need for giving you the background to the topic, or for helping you understand lecture materials in your first and second years. They may not, however, provide enough detail for the later parts of your course, or for researching material for coursework.
- Textbooks are often written by a small team of authors, who will not be specialists in all of the topics covered in the textbook. As a result, the quality of the coverage of the different topics may vary. For some texts, the book is prepared by editors, with subject specialists writing the individual sections. These are more likely to have a more even quality of coverage of the topics but may be pitched at a more advanced level.
- Textbooks do go out of date, and this is likely to happen faster than for your school or college textbooks. It can take several years for a textbook to be written and to be published and then there may be a few years more before the book is revised and a further edition published. Whilst this may not cause problems for their usage for basic reading, (since the key principles of a topic are unlikely to change significantly) if you require more specialist information, then it is more important that the book you are using is up to date. So, when you are scanning the shelves of your library for textbooks, make sure that the books you select are not more than two or three years old or that you purchase the latest edition.

5.1.2 Specialist topic book or monographs

As their name suggests these are books that focus on a specific topic within the subject. These usually take three forms.

1. An advanced textbook dealing specifically with a fairly narrow topic, such as liquid crystals. This will clearly be more detailed in approach than the broader textbook on organic chemistry but will be written in the same style. It will almost certainly have been written by a specialist, or group of specialists, within the field.
2. A monograph usually takes the form of a long article or a short book on a specialist topic. This will typically be narrower in focus and at a more specialist level than the advanced textbook.
3. At the research level, there are often books that are published that are based on the papers presented at a specifically themed conference. As with journal research papers, the individual papers may be specialized, being written for a readership of research scientists within the field. However, the grouping together of a whole series of papers on a specific theme may well provide you with a valuable starting point for your research for an essay topic. These books also often contain useful overviews of groups of papers in review form.

As with using textbooks, you need to make sure that the information you are looking at is up to date; indeed, the more specialized the book or research paper, the faster it is likely to go out of date. All of these, however, make valuable resources for advanced reading around a topic to increase your depth of knowledge and also often make very useful starting points for research for essays or similar assignments.

5.1.3 General science journals

There are several general science journals, such as *Chemistry World*, *New Scientist*, and *Scientific American*, that publish review-type articles aimed at the general interest reader with a scientific background. These types of journal can be very useful in providing overviews of a specific topic and also for keeping you up to date with wider developments in science as a whole.

5.1.4 Specialist review articles

Many research journals include review articles as well as the papers describing the results of novel research. There are also numerous journals dedicated to reviews in chemistry, such as 'Reviews in ...' series or the 'Progress in ...' series.

Recent review articles are a gold mine for students seeking information about a specific topic. These articles are typically written by a well-known specialist in the field who brings together, in a single article, a synthesis of a large number of individual research papers. The approach of these articles is usually designed to enable someone with some knowledge of the field, but not necessarily at a very detailed level, to derive a picture of the current ideas and developments within that field without having to read all the individual research papers.

Review articles are very useful for providing more detailed information about a topic, for example for supplementing lecture material, particularly at third- and fourth-year levels. They also often make very useful starting points for research for coursework assignments, both because of the overview they provide and because of the references made to recent research work in the field. Your lecturers will almost certainly include some review articles in the reading lists for your courses, particularly in the later years of your degree programme.

5.1.5 **Research papers**

Research papers published in academic journals are the dynamic bed rock on which current knowledge is based. These papers are written by scientists describing their latest discoveries or developments within their field of research. In any honours degree programme, there will be an expectation that you will:

- read recent research papers in order to supplement the lecture material;
- use them in order to prepare your coursework assignments;
- use the information you have acquired from reading research papers to support the arguments you present in examination essays (this will be particularly true for finals examinations).

As with the review articles, your lecturers will certainly include recommended research papers (some of which they have probably written themselves!) in the reading lists they give you to supplement the material covered in the lectures, especially at the final-year level. The fact that someone who is an expert in the field has recommended a given paper suggests that the paper has merit as a piece of scientific writing and that the findings are likely to be reliable. In the same context, papers for publication in scientific journals are subjected to a peer-review process: in this process, reviewers (usually two), who are acknowledged experts in their field, are asked to read the paper and to comment on its scientific value, to rate its importance in the field and suggest any amendments they think should be made. Whilst this is clearly not a fool-proof process, and is certainly not an excuse for you to stop thinking critically when reading, it does give these papers a stamp of authority and reliability.

Research papers are the most specialized of all the standard sources of information and can therefore be quite intimidating and hard to tackle: for example, they will probably use terminologies and be based on ideas that you do not understand very well. Research papers are usually in a highly stylized format, presenting information in a way that you are probably not accustomed to. Finally, it must also be admitted that they don't always make the most interesting bed-time reading! So, if you are going to invest time and effort on reading research papers, you need to be very clear about your reading goals so that you get the most benefit from your input.

The structure of research papers

Most research papers are written in a standard format, comprising seven main elements (see also Chapter 8, *Writing practical reports*).

- *Title.* The title is a short descriptor of the paper, as such it will give you a good idea of whether the subject matter is relevant to your needs.
- *Abstract.* The abstract is usually about 200 words in length and provides a short summary of the paper. In particular this will include the aims of the research, the key results, and a brief commentary about the conclusions. It may be that this is all you need to read in order to glean the key points, for example to support the argument in an essay.
- *Introduction.* The introduction is the section of the paper where the authors set out the background to the study. This is often very useful both for helping you to understand the basis of the topic and the reason why the researchers undertook the investigation they are describing. As such it will normally include a useful summary of the preceding research and an analysis of the questions still to be answered. It usually culminates in a setting out of the aims of the study.
- *Methods.* The methods section describes the way in which the research was undertaken, including details of the experimental protocols and the procedures for analysing the data. Unlike the type of report that you might write (see Chapter 8) the protocols are often not described in full but rather refer to previous papers in which they have been described, for example:

  The ruthenium catalyst was synthesized and activated according to the method described by Smith *et al.*[1]

  This form of summary information can be fine for people working in the field who already have a good knowledge of the literature. For the student trying to understand the paper, however, this can be a nuisance because it means further searching out of papers in order to find the actual description of the method, particularly if, as in this case, the method is an old but standard technique.
- *Results.* In the results section the authors will describe the results that they have obtained from their research and will also display them in different formats, for example as graphs, tables, or other illustrations. Where quantitative data is being presented this will also be analysed using statistical techniques. Note that this section is descriptive and does not normally include any discussion of the results or reference to the findings of other researchers.
- *Discussion.* The discussion section, as its name suggests, is where the authors interpret the findings of their research both in terms of the specific experiments they have

undertaken and also in the context of the current hypotheses in the field, as developed by other researchers. This section should link back to the introduction where the aims of the project were set out. Again, this section makes very useful reading as it links what has been reported in the paper to the existing research literature.

- *References.* This comprises a list of all the research papers, etc. that were referred to in the text of the paper. The reference list is often very useful in providing a source of further reading to help you fill in the background of your topic.

Try this: Find information by chasing back through the literature

In their paper 'Chiral Pd aqua complex-catalysed asymmetric C–C bond-forming reactions: a Brønsted acid–base cooperative system' (*Chem. Commun.*, 2009, 5787–5798) Mikiko Sodeoka and Yoshitaka Hamashima refer to their use of the Michael reaction with enones as a model reaction. Use the references in this paper to find this reaction and sketch the reaction scheme.

Reading a research paper

From the reading list supplied by your lecturer and your own searching of databases, you will probably have collected a frighteningly long list of research articles on a topic. Unfortunately (and contrary to some popular opinion) the acts of downloading a paper or getting a copy from the library do not also transfer the contents of the paper to your brain. You actually have to read the paper in order to extract the information from it. You also know that there is no way in which you have enough time to read all these papers.

So what should you do? This is where you need to be selective in your searching (see Section 5.2) and think about your reading goals to decide what you really want to get from your reading. For example, is it some specific, factual information to add in to an essay? Or is it a detailed understanding of the topic as the background to a final-year research project? It may be that just reading the abstract will suffice, or you might need to read the whole paper. These different approaches can be evaluated by careful consideration of your reading goals.

Goal 1

I am reading papers to find specific information for an essay or practical report.

Action 1

In this case, you are looking for specific information and do not necessarily need to read the whole paper. Check that the title appears relevant and then read the abstract: this should summarize the key points of the paper and may by itself be sufficient for what you need, particularly if the main use of the reference is to confirm a minor point in the essay. However, if the information you require is more detailed, then you will need to go deeper into the paper and use some of the strategies described below.

Goal 2

I am reading papers to support the lecture material on a module.

Action 2

For this type of reading, you may often begin from a reading list that you have been given along with your lectures. Therefore, you can start by selecting the papers from that list that appear directly relevant to the topic you are reading up on. As with Goal 1, you may be able to be quite selective in the extent of your reading, particularly if there is a specific point that you want to know more about. Alternatively, you may also wish to read the papers fully in order to appreciate the wider aspects of the topic. Try to make sure, before you start reading, that you are clear in your own mind why you are reading that paper and what you want to get from it.

Goal 3

I am preparing the critical literature review for my final-year project.

Action 3

For this type of exercise you will probably need to read the paper carefully, especially if you are comparing the hypotheses put forward by different authors. As with all your reading, when you get to a section relevant to your reading goal you should make careful, referenced notes about what you have learnt from each paper. This is particularly true when engaged in a long-term piece of work such as your final year project because you may well need to be able to go back to the sources to check, or add to, what you have written.

So, how do you find this information if it is not taken from a reading list? The most common approach is to use internet-based searches, which we describe below.

5.2 Using the internet

Anyone who has looked on the internet will be aware that there is a vast, bewildering wealth of information available at the click of a mouse button. Some of that information is accurate, up to date, authoritative, and really useful. A lot of what you may unearth, however, is none of those things. So, while online searches can be very productive, you need to have a clear strategy for directing your search, for filtering the material that is available and for being able to evaluate the value of that information.

Let us say that you have been set an assignment on the applications of organometallic catalysts and you need to find out some recent information on the topic. If you need to start by finding out basic information, then it is probably best to start with lecture notes and handouts, progressing on to text books and reading lists as described above. When you have exhausted these, you need to undertake your own specific searches.

There are two main starting points: you can use a search engine to explore the whole of the web, or you can look specifically for articles in academic journals. In this exercise, we will explore both approaches for comparison.

5.2.1 Academic journals

When you are searching for specialist scientific information, then academic journals should be one of your main sources of information. Almost all universities and colleges subscribe to electronic journals and also to databases that search the academic literature. Through your own library's catalogue, you should be able to find what journals you will have direct access to via an online link. The library catalogue should also provide links for you to use databases for your search. Examples of such databases that are commonly employed by chemists are Web of Science, Royal Society of Chemistry, American Chemical Society, Beilstein, and Gmelin. Different databases cover different ranges of journals and so they vary in size and in terms of the number of journals included. Some, such as Web of Science, are relatively broad covering all areas of science. You will need to select which database you use; it may be advisable to seek direction from a university librarian who will have a good knowledge of the databases your university subscribes to and the range of topics that they cover.

These databases allow very sophisticated searching strategies using a range of criteria. The most common strategy is likely to be based on a set of keywords that you think are core to the topic. If you have been given the name of a specific researcher, you could also search using the author's name. The advanced techniques of searching are beyond the scope of this book, and you should consult with an adviser from your library, but we will look at some simple strategies for searching, which will probably be adequate for much of what you want to do.

You need to find out about the applications of organometallic catalysts so you could start with a crude search in Web of Science and simply enter the words 'organometallic catalysts'. This comes up with over 2300 research papers. This is more than you probably have time to scan the titles, let alone read! So how can you refine the search strategy to pull out the articles that will be of value? There are several basic options here:

- choose specific types of article;
- select only more recent publications;
- refine your search terms.

See Figure 5.1 for an example.

Choose the type of article

Most search engines allow you to specify what type of article you would like to select. You can do this either by specifying what types you want to *select* or what types you want to *exclude*. For example, if we only specify review articles, then the searches come up with

FIGURE 5.1 Example of page 1 of the results from a Web of Science search, searching on the term: organometallic catalyst.

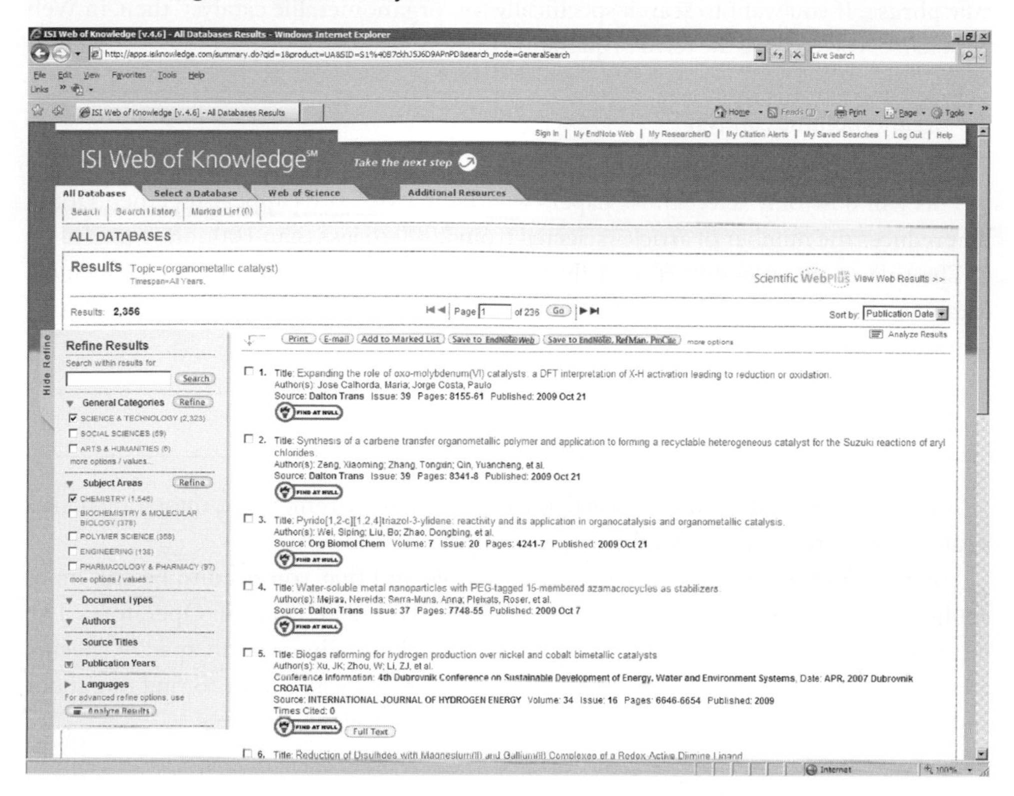

about 200 articles, but not all of these will be directly relevant because of the breadth of the search terms used (see below). Furthermore, there is a risk that you will miss out on some useful research papers that have not been covered by a review article.

Select only more recent publications

Here you can choose what years' publications you want to consider. This allows you, for example, to narrow down to a single year if you wish. A rule of thumb for your assignment might be to narrow your search down to the last five years as a starting point. If we select only the last three years, then the number of articles from Web of Science drops from over 2000 to around 200. That's still too many but it is a considerable improvement. Again there is a risk that you may exclude some older, but very useful papers.

Refine your search terms

'Organometallic catalyst' is quite a broad term to use for searching, so you could consider refining your strategy. If you enter the term 'organometallic catalyst' into the search box, then the database will search for all articles that include the terms 'organometallic' and 'catalyst' but this does not mean that the words will necessarily appear together as a specific phrase. If you want to search specifically for 'organometallic catalyst' then, in Web of Science, you enclose the term in quotation marks:

"organometallic catalyst"

This will now only select those papers where the two words appear together. Doing that reduces the number of articles selected from 2000 to less than 100.

The title of the assignment actually referred to the 'applications of organometallic catalysts'. So a further refinement would be to include the word 'applications', so your search term becomes:

"applications of organometallic catalyst"

This now reduces the number of articles to zero but if we remove the quotation marks to allow the words in any order then we get 100 articles.

You may find that not all searches are as simple and that you need to be still more sophisticated. A further refinement is to use what are called 'Boolean Operators'. The most common of these are AND, NOT, and OR. These allow you to specify specific combinations of search terms. For example, if you specifically want to find out about ruthenium organometallic catalysts you could enter the following string:

"organometallic catalyst" AND ruthenium

This will now return a list of articles where the words organometallic catalyst occur together and the word ruthenium also occurs. However, if you wanted to find out about any work that had been carried out on metals other than ruthenium, you would enter the search string:

"organometallic catalyst" NOT ruthenium

Through these types of search operators, you can create very specific criteria for what articles you want.

Each time you undertake a search, you should look at the number of articles that the search has returned and scan the first few titles to check that they appear relevant. With a broad search you are likely to find a large number of articles, many of which are not directly relevant. As you refine your search, for example by selecting specific years or by using more specific search terms, then the number of articles should progressively reduce and should become more relevant, but there is a compromise as you may also start to exclude some articles that would be useful.

Try this: Try to refine a search on Web of Science

Arsenicals are arsenic-containing compounds that are used as pharmaceuticals. Starting with the term 'arsenicals' use Web of Science to narrow down your search to find recent articles describing the synthesis of sodium-containing arsenicals.

5.2.2 Web search engines

The major web search engines such as Google, Google Scholar, Yahoo!, AltaVista, etc. are familiar to most people and they can be used to scan literally billions of pages of web-based information. They are very effective at producing long lists of websites related to your inquiry, but this means you have to be very selective both in terms of finding material that is directly relevant and then evaluating the reliability of that material: don't forget anyone can host a site and put information onto the web and there is no guarantee that the information has been checked for accuracy.

Researching for our essay on the applications of organometallic catalysts using search engines, we can adopt the same procedures as when using the academic journal databases. If we start simply by entering the key words:

---

applications organometallic catalysts

Google returned 363 000 results, Yahoo and Alta Vista both returned 176 000, and all three engines ranked Wikipedia at or near the top of the list. The ranking of the list, however, does not necessarily indicate the direct relevance of the information to be found on the site (see Box 5.1).

BOX 5.1 Web search engines: a very brief introduction.

When we are looking for information about organometallic catalysis and enter those words into the search box of a search engine, the search engine will return sites that contain the words 'organometallic catalysis', but the words will not necessarily be together and in that order unless you have used an advanced search strategy.

The order of the listing of the sites is also determined by a ranking process. This process ranks pages on a scale of 1–10 on the basis of the number of links there are to that specific site. The importance of each of those links is also scaled on the basis of the ranking of their sites. So the more links there are to a given web page and the higher the ranking of each of those sites, then the higher that web page will appear on the search engine's listing.

This means that, although the first pages in the listing will be popular, they are not necessarily the most relevant pages and you may need to scan quite a way through the listing to pick up the most relevant sites.

Although there are clearly too many sites to check more than just the first few, a scientific topic such as homogeneous catalysis tends to come up with a significant number of research papers from academic journals and links to other academic sites such as universities in the first listings, so these are likely to be reliable sources for our essay.

If the title is on a more topical subject, the results are very different. For example, let us we consider the 'global warming' as a title. Using the keywords 'global' and 'warming', the results from Google returned 29 200 000 results.

It may be that, for the piece of work you are researching, you do need a range of sources of information and not just research articles. You can therefore use the broad search engines and maybe refine your search using the advanced search techniques available. As with the academic databases referred to earlier, these advanced search techniques allow you to specify search criteria, such as an exact phrase or without specific words, etc.

If you want to search for academic articles only, then the search engines now include subsections that will do that. An example is Google Scholar – a search engine that searches peer-reviewed papers, books and conference proceedings. Although the range of journals that can be accessed by these types of searches is not as extensive as for the specifically academic engines, such as Web of Science, there has been a rapid expansion and search engines such as Google Scholar do offer a good alternative. Google Scholar found 131 000 articles for our search for applications of organometallic catalysts.

How reliable is a website?

As we have seen, there are many sites providing information on specific topics and some are more reliable than others. So how do you know what to use? From a purely academic standpoint, the scientific journals, with their peer-reviewed papers offer the most reliable sources of information but there are many other useful sites as well. The simplest questions to ask yourself about each site are: who, or what organization wrote the material and how old is it?

You can quickly answer these questions by looking at the site address in the listing. Each site is identifiable by its URL (Universal Resource Locator). Examples of useful sites would include:

- *University sites*. These can include course materials from the university's undergraduate programmes, which can be very useful as an additional resource, as well as links to the publications and research outputs of their academic departments. Sites for higher education institutes are identifiable from the URL, which will contain the code ac (academic) or edu (educational). Thus, universities in the UK all have an address that takes the form: www.[universityname].ac.uk. Those in the USA have the address www.[universityname].edu and for Australia they are www.[universityname].edu.au. Be aware though that undergraduate students may have placed resources on their departmental website and these should be treated with caution.
- *Government sites*. These can provide information at a range of levels, particularly in terms of policy or public advice. Examples here would be the Food Standards Agency

or the Environment Agency. Most government sites in the UK have the address form: www.[name].gov.uk.

- *Professional bodies and learned societies.* The Royal Society of Chemistry is the professional body for chemists in the UK. You may also come across the Society of Biology and the Institute of Physics. These sites usually have addresses as: www.[name].org
- *Media organizations.* Many of these organizations, such as the BBC, *The Times*, *The Guardian*, and *The Daily Telegraph*, etc. are very useful for providing information about topical science stories. Remember, however, that the story is usually written up by a journalist who may have a limited knowledge of the topic and is aimed at the general public rather than someone looking for detailed, scientific information. Media organizations are also in the business of selling news stories and, therefore, they may focus on the more sensational aspects of a topic.
- *Commercial organizations.* Many commercial organizations have very useful websites but it is important to remember that they also want to sell a product. Going back to our search on global warming, there were several sites listed that were hosted by commercial organizations selling products to assist reducing CO_2 emissions. These carried quite a lot of useful educational information. However, they also carried a lot of commercial information setting out the benefits of their specific product. Whilst the information published by such sites has to comply with the requirements of the Advertising Standards Agency, they do not have to be objective in their approach: for example, while they are extolling the benefits of their own product, they are under no obligation to inform you that another company's product may actually be better!
- *Specific societies and organizations.* There is a huge variety of organizations and groups who have websites and the quality of these varies greatly. In the area of chemistry these include the Chemical Industries Associated, the Association of British Pharmaceutical Industries, the LGC, and numerous others. These sites often carry a significant amount of background information. However, these organizations also often have very focused agendas and therefore the information provided may be selective or interpreted in very specific ways, so when you are reading, you need to be aware of that focus and to maintain your critical approach.

Mention must also be made of one of the most used sources of information on the web, Wikipedia. Wikipedia is ostensibly an online encyclopaedia but it differs from all other encyclopaedias in that anyone can add or delete information from specific entries. This means that, whilst much of the information present may have been written by experts and should be accurate and current, an entry may also have been written by someone with very limited knowledge and may contain major inaccuracies. Unless you already know about the topic, you will not know whether the content is accurate or not. Wikipedia may have uses, for example for reminding yourself quickly about a topic, but it is certainly not advisable to use it if you are learning about a topic from scratch or researching for an essay. Your lecturers are unlikely to be impressed if the reading list for your assignment contains articles from Wikipedia.

5.3 Note-making strategies

It's past midnight and you are finally in the process of writing your essay. It's due in for tomorrow morning but that's okay because you have done all the research and read the papers, you have your plan in front of you and the writing is going well. Then you hit a problem: you just need to describe the key experiment that supported the argument you are putting forward. You know you read up on it but you can't quite remember the details. It's somewhere in that pile of review articles but you can't remember which one and you don't have any notes to jog your memory…

Situations like this are very frustrating but they can easily be avoided if you are systematic in making notes when you read up about a topic. Many students only think of note making in the context of lectures but it is just as important in terms of your reading around the subject. We can divide note making up into two processes: the first is generic and comprises the things you should be doing irrespective of why you are making the notes. This includes making sure that the information you are using is properly referenced so that you avoid plagiarism (Chapter 10, *Avoiding plagiarism*). The second process varies depending on what you want to use the notes for, e.g. developing understanding of a topic, revision purposes, or information for a coursework assignment.

In the following sections, we consider each of these in turn.

5.3.1 Generic note making

To ensure that your notes are as useful as possible and that your note making is done effectively, you must always record the source and make sure that your notes are notes.

Record the source

Always make sure that you record the source of the information. You never know when you may need to find it again: particularly when writing your essay in the early hours of the morning... the more details you give, the easier you will make life for yourself later on (Figure 5.2). For example, just jotting down the title of a book is of limited help, particularly if it is several hundred pages long! Don't forget, also that if you are note taking for an assignment, you will need the full reference for your reference list.

Notes are notes!

You should always aim to note the information in your own words: just copying a paragraph out, or photocopying it, is not note making. The process of reading and then summarizing the information in your own words is a very useful exercise because:

- as an active process, it will help you retain the information in your memory;
- it also tests your understanding of the information;
- it makes it faster for you to retrieve the key information at a later date, e.g. when revising;
- it makes you less likely to be tempted to plagiarize.

FIGURE 5.2 Reference details for notes from a book and a research paper. Note that it is okay to use abbreviations in the details since you can go back to the full reference if you need it for your assignment.

Atkins, Overton et al (2010) Inorganic Chemistry.
Chap 26, p697
Cycle for Wilkinson's catalyst ...

Characteristic reactions of group 9 transition metal compounds in organic synthesis, Omae (2009) App. Organomet. Chem 2009, 23, 91–107
Page 98 for structure of Rh catalyst

Have a look at extract from the sample text in Figure 5.3 and then the notes in Figure 5.4 taken from the sample text. Identify some of the techniques used to set out the notes. These include:

- use of highlighting to emphasize the topic;
- leaving space to keep the notes clear;
- leaving a wide margin, so you can add additional notes or explanatory comments later on;
- the source of the material (including page number) is given at the start of the notes;
- lines are used to indicate linkages of ideas and the division of the types of disorder into two groupings;
- the writing is in a very abbreviated style: there is no need to write in full sentences or include words like 'the', etc.

5.3.2 Focused note making

As with your reading goals, the way in which you make notes will depend to some extent on the reason why you are making the notes. Some possible examples are:

- to aid with revision;
- as a tool to help with learning about a new topic;
- to aid understanding of a topic;
- as part of the research for a coursework assignment.

Assuming that you will be adopting the generic approaches described above, how do these goals differ in terms of note making? One way of looking at them is to look at them in terms of your requirements. For example, the first two goals are likely to be fairly broad

FIGURE 5.3 From Atkins, Overton, Rourke, Weller, Armstrong 2010. *Inorganic Chemistry*, Oxford: Oxford University Press.

26.2 **Homogeneous and heterogeneous catalysts**

Key points: Homogeneous catalysts are present in the same phase as the reagents, and are often well defined; heterogeneous catalysts are present in a different phase from the reagents.

Catalysts are classified as homogeneous if they are present in the same phase as the reagents; this normally means that they are present as solutes in liquid reaction mixtures. Catalysts are heterogeneous if they are present in a different phase from that of the reactants; this normally means that they are present as solids with the reactants present either as gases or in solution. Both types of catalysis are discussed in this chapter and will be seen to be fundamentally similar.

From a practical standpoint, homogeneous catalysis is attractive because it is often highly selective towards the formation of a desired product. In large-scale industrial processes, homogeneous catalysts are preferred for exothermic reactions because it is easier to dissipate heat from a solution than from the solid bed of a heterogeneous catalyst. In principle, every homogeneous catalyst molecule in solution is accessible to reagents, potentially leading to very high activities. It should also be borne in mind that the mechanism of homogeneous catalysis is more accessible to detailed investigation than that of heterogeneous catalysis as species in solution are often easier to characterize than those on a surface and because the interpretation of rate data is frequently easier. The major disadvantage of homogeneous catalysts is that a separation step is required.

Heterogeneous catalysts are used very extensively in industry and have a much greater economic impact than homogeneous catalysts. One attractive feature is that many of these solid catalysts are robust at high temperatures and therefore tolerate a wide range of operating conditions. Reactions are faster at high temperatures, so at high temperatures solid catalysts generally produce higher outputs for a given amount of catalyst and reaction time than homogeneous catalysts operating at lower temperatures in solutions. Another reason for their widespread use is that extra steps are not needed to separate the product from the catalyst, resulting in efficient and more environmentally friendly processes. Typically, gaseous or liquid reactants enter a tubular reactor at one end, pass over a bed of the catalyst, and products are collected at the other end. This same simplicity of design applies to the catalytic converter used to oxidize CO and hydrocarbons and reduce nitrogen oxides in automobile exhausts (Fig. 26.4), see also Box 26.1.

FIGURE 5.4 Example of notes taken from a piece of text.

Inorganic Chem, 2010, Atkins et al, ch26

Homo vs hetero catalysis

P 694

Single phase	different phases
Usual liquid	usually solid
Highly selective	economics
Highly active	robust
Diss heat	fast Xns
Investigate mechanisms	environm friendly
Need to sep products from catalyst	no separation

in coverage of a topic, whereas the last two are likely to be more focused on a specific aspect of a topic.

Note making to aid with revision

You are preparing for an exam and have a large volume of notes from lectures, some background notes from reading textbooks, and perhaps some notes from research papers. It is a very useful exercise to bring these notes together so that they form a coherent theme that you can read through, rather than chunks of information dotted around. As well as making the notes easier to revise from, the process of actually undertaking such an exercise also helps you learn. This topic is addressed in more detail in Chapter 14, *Getting the most out of revision.*

Note making as a tool to help with learning about a new topic

It may be that you have to research a topic from scratch rather than using note making as a supplement for your lecture notes or handouts. In this case, the best approach is to build up your notes in stages, rather like the revision notes. To start with you need to get an overview of the topic, for example by reading and making notes from a general textbook (see Section 5.1.1, goal 1). Then go into the topic more deeply using a review article or research papers to focus on specific aspects. Finally, produce a set of composite notes by combining the overview notes with the more detailed ones. By the time you have done that, you should have a useful set of notes but should also have gained a good appreciation of the topic. Furthermore, when you come to produce the composite notes, it will also become clear to you if there are still areas that you do not understand.

Note making to aid understanding of a topic

This task is a more focused note-making activity since you are likely to be concerned with a small aspect of a topic that you do not understand. In this case, your reading will be very goal directed (Section 5.1.1, goal 2) and you only need to make short notes to supplement your current notes from your lectures. In this context, it is very helpful if you have left large margins or space in your lecture notes (Chapter 2, *Making the most of lectures*) because then you may be able to insert the additional information within the original notes rather than having to write it separately. Whilst amalgamating notes from different sources is a very useful exercise, as discussed above, it is not necessarily such a good use of your time if there are only very small sections of additional information that need to be included.

Note making as part of the research for a coursework assignment

As part of the preparation for your assignment, you should have drawn up a list of specific topics that you need to address (see, for example, the section on *Make a plan* in Chapter 7, *Writing essays and assignments*). Under each of those headings, you can now set out the key pieces of information from your reading, which will supplement other sources such as your lectures. As for understanding the topic (above), each point is likely to be quite brief and you will probably only need to record three pieces of information:

- the factual information itself;
- a summary of the experimental approach used to find the information;
- the full citation for the reference (see Section 5.4).

5.4 Citations and references

When you are writing or presenting any piece of work, it is very important that you acknowledge the sources of the information and ideas that you are presenting. If you present an idea or a piece of information without acknowledging a specific source, then you are effectively claiming ownership of that idea or information. Provided it is your idea then that is okay: for example it could be your interpretation of the results obtained in a practical class or from your research project. The failure to acknowledge sources, however, is one of the most common reasons for being accused of plagiarism (see Chapter 10, *Avoiding plagiarism*).

In this section we will look at a standard way of referencing and discuss when and how you should do it.

5.4.1 Defining the terms

The process of acknowledging sources sometimes seems more complicated than it need be because of the different terms used to describe what you should do. We will work through some examples but it is useful to explain some of the terms first.

Referencing is the name given to the process of acknowledging any material in the form of ideas, specific information, drawings or illustrations, experimental methods, computer programs, etc. that has come from someone else's work. Normally this is work that has been published in some form, for example as a research paper or book.

Citations are the specific acknowledgements to other people made, for example in the text of your writing. Normally, these will be the name of the person and the date when their material was published: see some of the examples below in the short section on *How to draft citations and reference lists* (p.66).

References are the list of sources that you actually referred to in your work. Every citation in the text should appear in the reference list and every source listed in the references should refer to a citation in the text.

Bibliographies are lists of all the sources that you made use of during your research. A bibliography may, therefore, include sources such as general textbooks that you used for background reading but from which you did not actually take specific material for your work. Some of the material listed in the bibliography may not, therefore, appear as a citation in the text. For most work at university level in chemistry, particularly in the later years of your course, you will be expected to produce a reference list rather than a bibliography.

5.4.2 Why should I cite other people's work

The main reason, as indicated in the introduction to this section, is so that someone reading your work will know where the ideas you are presenting came from and will be able to identify which ideas or information that you are presenting are specifically your own. The references are also very helpful to the reader who needs to know more about the background to the topic: he or she can go to the articles you have cited to find out more detailed information. When you are preparing coursework, the list of references will show your tutor what materials you have used and he or she may be able to give you feedback about the suitability of the materials and your search strategies.

5.4.3 Do I have to cite everything?

The general principle is that anything that you include in your work that is taken from someone else must be acknowledged. This is true just as much if you are quoting someone word for word, or if you are taking their ideas and putting them in your own words. Having said that, there is a level of information that is now accepted knowledge and for which it is not necessary to provide a citation. For example, if you write the statements:

Carbon dioxide is implicated in global warming.
Radon is radioactive.
Haemoglobin is the respiratory pigment involved in oxygen transport in humans.

These are accepted facts and there is no need to cite the source of the information. Indeed, the derivation of such information is often unclear anyway. To get a feel for how much you should cite, one of the best things you can do is think about it when you are doing your background reading: when you read a review article or a research paper, make [illegible] which they make specific citations. Have a look at the guidelines below as a framework for knowing when to cite your sources.

You should always cite the source when:

- you are quoting word for word from a source, whether it is a textbook, research paper, lecture, or film. Any such quotation must also be written in quotation marks (see below);

- you are copying, or taking elements from, a figure from a piece of published work;
- you are presenting information or ideas that are critical to the arguments you are making;
- you are presenting any ideas that may be controversial, for example someone's theory about a specific process;
- you are presenting information based on specific experimental or observational evidence.

If in doubt, then acknowledge the source.

5.4.4 How to draft citations and reference lists

If you look through different academic journals and, particularly if you look at publications in different subject areas, you will see that the citations and reference lists are formulated in a variety of ways. There are several standard styles for referencing such as Harvard, Vancouver, Chicago, and so on. Although the styles are different for these, the basic information is always the same. In this context, Harvard and Vancouver systems are the most common formats for citations and reference lists in chemistry.

In the Vancouver or numeric system the reference is indicated by a number in the text, either as [1], or as a superscript[1]. In the reference section, the references are listed in numerical order, which can make them easier to find when reading a long paper or article. In the reference list you must also include the author (or authors') initial(s), the year it was published, the title of the article and the journal or book it comes from, the volume number, and the page number. In chemistry it is most common to use the numerical method of referencing. In the Harvard system, the citation in the text simply comprises the name(s) of the author(s) and the date. The list of references is alphabetical by author name.

The Harvard method is sometimes used in chemistry for discursive writing as in essays (and in this book). Note that the order in which the information is presented varies between the two methods. In the examples below we will show both methods and run through the most common variations for citations and references. Note that, even within the Harvard and Vancouver systems, there are differences in presentation. The formats given below are based on the conventions used by the Royal Society of Chemistry and the American Chemical Society but you should check with your lecturers whether your department recommends any specific styles.

Referencing journal articles

For much of your study, articles published in scientific journals will represent the major source of information. Note that the formats for referencing given below apply to all journal articles, even if you downloaded the article from the journal website.

If there are up to three authors for a given publication, then all the authors' names are given.

Citation:

Using the Vancouver system the citation would look like this:

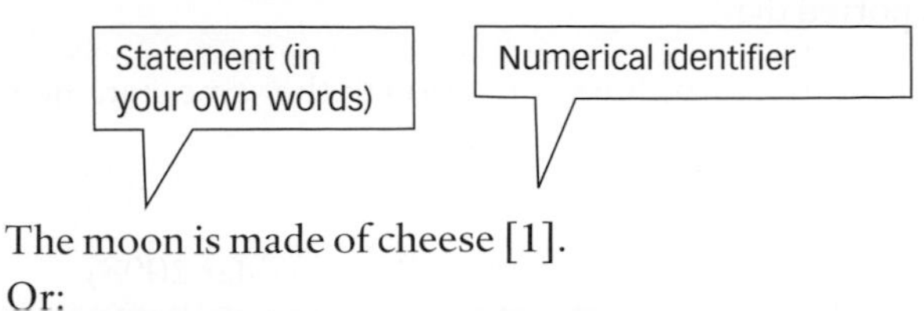

The moon is made of cheese [1].
Or:
Brie and Camembert[1] reported that the moon is made of cheese.

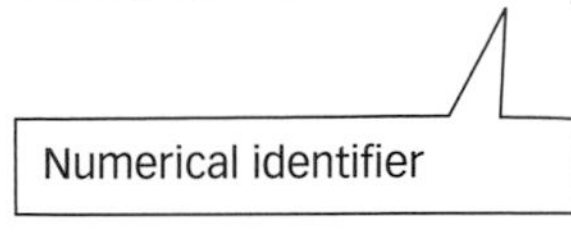

Using the Harvard system it would look like this:

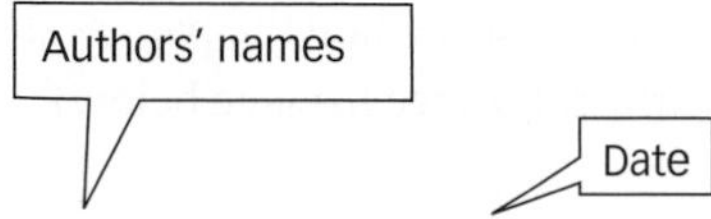

The moon is made of cheese (Brie and Camembert 2007).
Or:
Brie and Camembert (2007) reported that the moon is made of cheese.

Reference:

The Vancouver style reference looks like this:

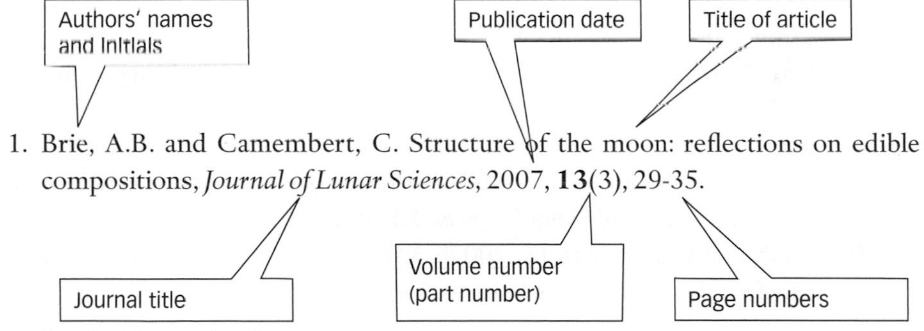

1. Brie, A.B. and Camembert, C. Structure of the moon: reflections on edible compositions, *Journal of Lunar Sciences*, 2007, **13**(3), 29-35.

Or using the Harvard style:

Brie, A.B. and Camembert, C., 2009, Structure of the moon: reflections on edible compositions. *Journal of Lunar Sciences*, **13**(3), 29-35.

If there are four or more authors, only the first name is given in the citation followed by *et al.* (this comes from the Latin expression *et alii*, meaning *and others*) but all the authors are listed in the reference.

Citation:

Brie *et al.*[2] reported that the moon is made of cheese.
Or:

The moon is made of cheese (Brie *et al.* 2007).

When giving a direct statement, as above, some authors prefer to us a phrase such as:

Brie and coworkers [2] reported that…

To cite two different papers by different authors in relation to the same statement:

Brie and Camembert [3] and Cheddar [4] reported that…
Or:
The moon is made of cheese (Brie and Camembert 2007; Cheddar 2008).

If you have more than two papers by different authors, the use of a numerical range, as in [4–8] or the listing of the authors and dates in brackets is usually preferable rather than a long list of names in the body of the text. Note that the publications are normally listed in chronological order.

To cite two papers by the same author(s) in different years, list the papers in chronological order. To cite two papers by the same author(s) in the same year only presents a problem in the Harvard system when the papers should be identified as 'a' or 'b' and this distinction added to the reference list (see below).

Citation:

The moon is made of cheese[9,10]
Or:
The moon is made of cheese (Brie and Camembert 2007a, b).

Reference:

9. Brie, A.B. and Camembert, C., Structure of the moon: reflections on edible compositions, *Journal of Lunar Sciences*, 2007, **13**(3), 29–35.
10. Brie, A.B. and Camembert, C., Variations in the density of cheese on the lunar surface, *Journal of Lunar Sciences*, 2007,**14**(1), 54–63.

Or:

Brie, A.B. and Camembert, C., 2007a, Structure of the moon: reflections on edible compositions. *Journal of Lunar Sciences*, **13**(3), 29–35.
Brie, A.B. and Camembert, C., 2007b, Variations in the density of cheese on the lunar surface. *Journal of Lunar Sciences*, **14**(1), 54–63.

The assumption so far is that these journal articles are articles that you have read: that is that these are **primary** sources. It may be that you wish to cite the work of a researcher but have not read the material in the original article, for example, if you have not be able to obtain a copy of the article. In this case, you may have found that the information you need is quoted in a more recent paper by another author, and so you can refer to this more recent paper. This is clearly more indirect than taking your information directly from the original research paper, and it is referred to as a **secondary** source. The use of such secondary sources is not ideal, but sometimes it is unavoidable, particularly when you cannot access the original article.

Referencing books

If you are wishing to cite material from a book, the citation in the text takes the same formats as for the journal articles, that is the name(s) of the author(s) and the date of publication. The reference list format varies according to the type of book. The basic format is as follows:

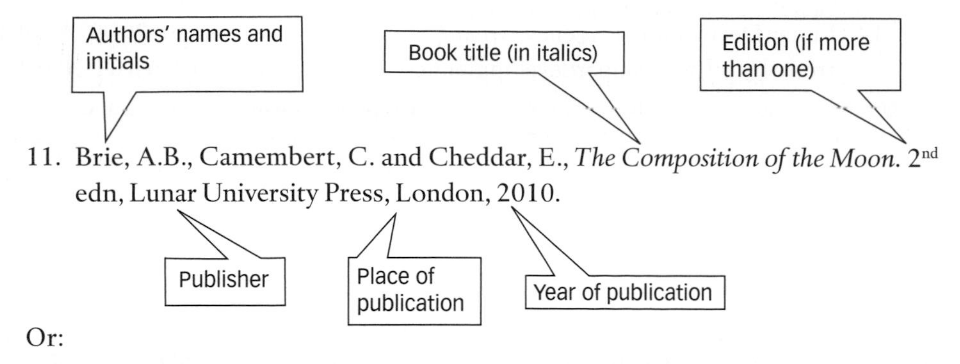

11. Brie, A.B., Camembert, C. and Cheddar, E., *The Composition of the Moon.* 2nd edn, Lunar University Press, London, 2010.

Or:

Brie, A.B., Camembert, C. and Cheddar, E., 2007, *The Composition of the Moon.* 2nd edn, London: Lunar University Press.

Many books are compositions of articles that have been edited by the main editors. In this case, the reference would be:

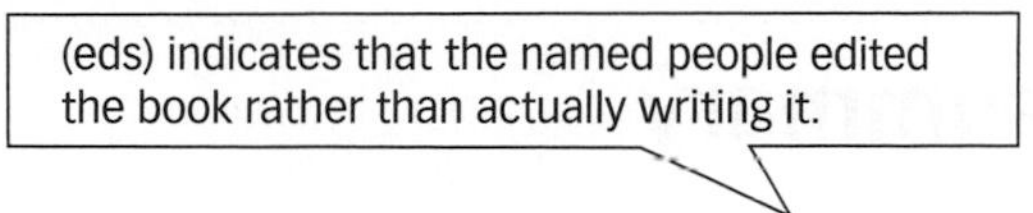

12. Brie, A.B., Camembert, C. and Cheddar, E. (ed.), *The Composition of the Moon,* 2nd edn, Lunar University Press, London, 2010.

In the case of such edited books, you may have only cited a specific chapter in the book that was written by another author. The format for such a reference is as follows:

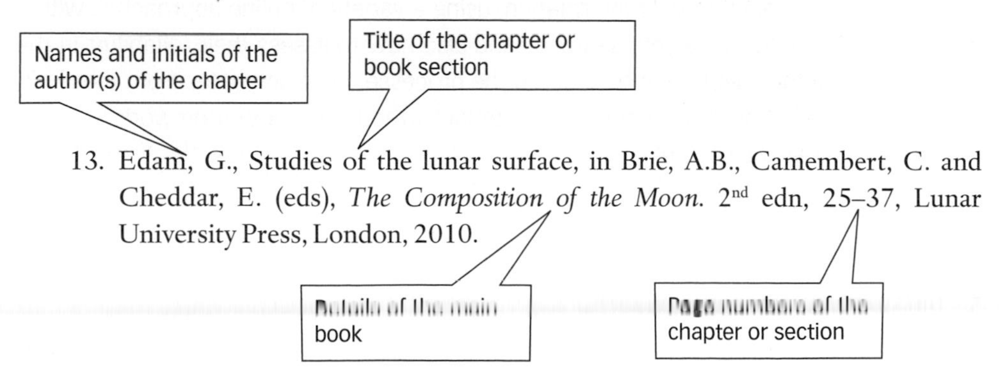

13. Edam, G., Studies of the lunar surface, in Brie, A.B., Camembert, C. and Cheddar, E. (eds), *The Composition of the Moon.* 2nd edn, 25–37, Lunar University Press, London, 2010.

Referenc websites

An increasing amount of information is available from websites and many such sites can be very valuable sources (though see Section 5.2). As with any other source of information, that source has to be acknowledged through a citation in the text and a reference in the reference list. The citation in the text is, as for the other types of source, author(s) name(s) and date of publication. The basic format for the reference list is as follows:

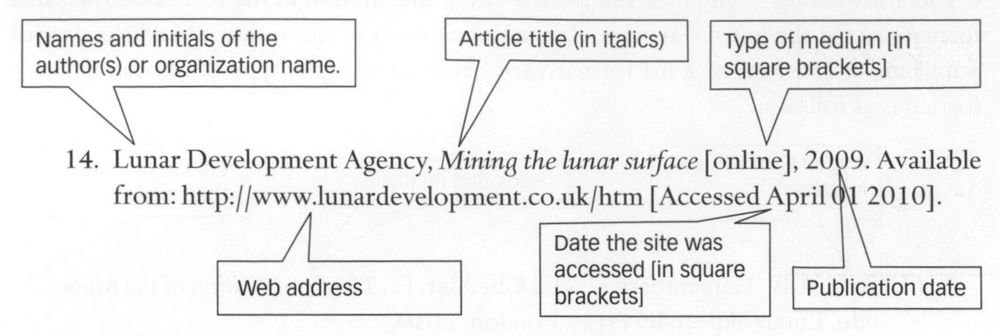

In this section, we have looked at the styles of citation and referencing for the most common sources of information that you might be using for your work. As stated before, there are different formats for presenting such information and here we have concentrated on the Harvard format, which is one of the standard forms used in chemistry. You should, however, ensure that you check with your university or college to see if they have a preferred style of referencing.

✱ Chapter summary

In this chapter we have looked at the range of ways of working with sources to find information. This started with looking at the different types of sources available to you, ranging from textbooks through to research papers, looking at the type of information each sources provides and the situations in which you might make use of them. We then reviewed the processes of searching for your information, using a variety of online approaches with a discussion of how to refine your searches and also how to assess their reliability. In the final sections of the chapter we considered the processes of taking notes from these different sources and how to reference the material correctly in your written work.

Remember that you should:

- understand the purpose of different types of publication;
- be able to carry out sensible literature searches;
- take notes for different purposes;
- select and use the appropriate style for referencing.

Chapter 6

Choosing the right writing style

Introduction

Writing for scientific purposes is often very different from the styles of writing we may use for other forms of communication, as it normally requires use of a formal, impersonal style. It is important to use words accurately and avoid slang, informal phrases, and generalities. When we are talking with each other, we can use gestures to add to the meaning of what we say and we can repeat things or state them differently if we realize the listener has not understood us. It is also very easy to be rather lazy in our expression because we can see whether our listener has understood, '...know what I mean?' When you are writing, you have only the written words to convey your meaning and you cannot ask the reader whether they have understood or not, so precision is very important.

6.1 Voice

The style for scientific writing is normally impersonal and so avoids the use of personal pronouns such as 'I' and 'we' and the expression of subjective viewpoints. It also commonly involves using the *passive* voice rather than the *active* voice. For example, an introduction to your essay could be written as:

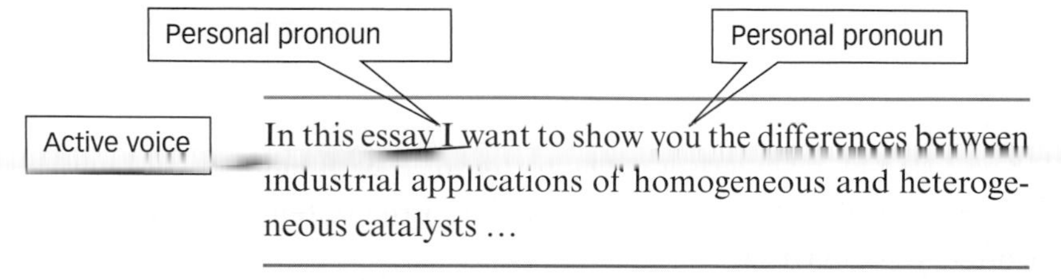

For a formal, scientific style it would be better to write:

Passive voice

> In this essay, comparisons will be drawn between industrial applications of homogeneous and heterogeneous catalysts .....

You should also avoid using contractions of words and expressing a subjective viewpoint. For example the conclusion to an essay might be written:

> The evidence I've given supports my feeling that...

This phrase contains the personal pronoun 'I' in the contraction 'I've'. You should avoid using any of the many contractions that appear in conversation such as 'I've', 'don't', 'couldn't' etc. Instead you should write them out in full: 'I have', 'do not', 'could not' (however, in this book we have chosen to use contractions because we feel that a more informal style is appropriate for this book, but it's (there we go again!) not appropriate for an academic essay).

The phrase also expresses a very personal, subjective viewpoint: 'my feeling'. In the formal voice, this could be written as:

> The evidence presented supports the argument that...

If it is important to express a personal summing up of the arguments presented, then you could write:

> On the basis of the evidence presented, I would conclude that...

The use of generalizations often clouds the meaning of what is written or indicates a lack of confidence on the part of the writer. For example, what does the following phrase actually mean?

> The value for iron concentration was quite often higher than............

The phrase 'quite often' has no real meaning as its interpretation depends entirely on the reader. Relative terms such as 'quite' or 'fairly', or phrases such as 'In general terms...' lack precision and should, wherever possible, be replaced with more precise phrases. Indeed, if you can, you should provide quantification of the statement, so rather than stating 'quite often' you can write:

> In 75% of samples the iron concentration was greater than...

As with the choice of words, precision in grammar is very important to ensure that our writing is clear and unambiguous. For example, the incorrect position of a comma in a sentence can change the whole meaning of that sentence as in the old joke about the panda that walks into a bar, eats something, pulls out a gun and shoots the barman before leaving the bar. When questioned, the panda pulls out a book and points to the description of the panda that states it '...eats, shoots and leaves'. Had the description been phrased correctly as '...eats shoots and leaves', the bartender would not have been shot! There are numerous, entire books covering the subject of grammar (for example *Eats, Shoots and Leaves*, by Lynne Truss) so in this brief section we will look at just a few of the key rules.

Having read this section you should now be well aware that we have not written this book in a scientific style!

6.2 Sentences and phrases

The fundamental unit of writing is the sentence. Sentences may be short. If too many short sentences follow on from each other, however, the style of writing can seem very abrupt and disjointed but overly long sentences, particularly if they contain several, inter-linked phrases, that might be associated with different points, can also be problematic because they can cause the reader to lose the thread of the argument being presented and give rise to the impression that the writer does not have a clear perspective of what he or she is trying to say.

The section you have just waded through is clearly badly written, and you may well not be sure what we were trying to say (as a short exercise, you could try rephrasing the paragraph so that the meaning is clear). So what is a sentence? At its most fundamental, a sentence comprises the following features:

- a noun, that is the subject of the sentence;
- a verb, that describes action;
- the beginning of the sentence is marked by a capital letter;
- the end of the sentence is marked by a full stop, a question mark, or an exclamation mark.

For example:

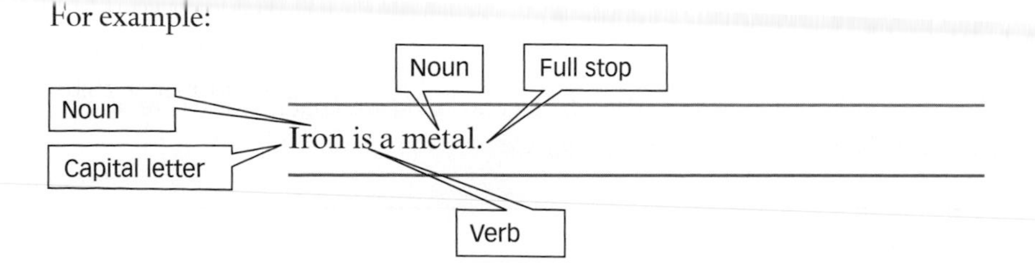

In fact, this sentence contains two nouns: 'Iron' and 'metal'. The noun 'Iron' is the **subject** of the sentence, and the noun 'metal' is used to give the reader more information about what iron is. Although it is very brief, it is a complete sentence, whereas:

> Iron is a metal that exhibits

is a phrase not a complete sentence. Although it contains a noun and a verb, the sentence is incomplete because the verb 'exhibits' needs to be qualified by a description of what property iron exhibits:

> Iron is a metal that exhibits several oxidation states.

is now a complete sentence. We may also include **adjectives** and **adverbs** in order to give greater detail to our descriptions to the reader. Adjectives are words used to describe a noun and adverbs are used to describe a verb. For example, the sentence below:

> The sample can be analysed by titration.

is a complete sentence but it does not give the reader very much information. By using adjectives and adverbs we can amplify the statement as follows:

Adjective — Adverb

> The aqueous sample can be analysed rapidly by titration.

In the sentence above, the adjective 'aqueous ' tells the reader what type of sample it is, and the adverb 'rapidly' indicates to the reader information about the analysis.

If we wish to add still more information about the procedure, we could add a separate sentence:

> The aqueous sample can be analysed rapidly by titration. It needs no further preparation.

BOX 6.1 Adjectives and adverbs

Adjectives are descriptive words used to describe a noun: 'a *brown* coat'. Adverbs are also descriptive and used to describe a verb: 'he ran *quickly*'.

In the second sentence, 'It' is the subject of the sentence but it refers back to the previous sentence describing the sample. As we mentioned earlier, too many short sentences make the writing disjointed, so we can link these two sentences together:

The aqueous sample can be analysed rapidly by titration and it needs no further preparation.

Here, the phrase 'needs no further preparation' is giving a more detailed description about the analysis of the sample. Note that this is now a **phrase**, and not a sentence, because there is no subject within the phrase itself. The phrase is joined to the rest of the sentence by the **conjunction** 'and'. Conjunctions are words used to link phrases together, the most common being 'and' and 'but'.

With the aid of conjunctions, we can link many phrases together and end up with very long complex sentences that become difficult for the reader to follow. On the other hand, as we have already mentioned, a series of very short sentences is also hard to read. So, how long should a sentence be? Unfortunately, there is no specific ruling on this but there are some rough guidelines you can follow. Most of the sentences in this section of the book are between 15 and 25 words in length. If your sentences are approaching 35 or 40 words in length (or more!) and comprise several phrases, you should think about dividing them up. Very brief sentences may be useful occasionally, for example to emphasize a point. If, however, you find you have written a series of sentences, each of which is less than ten words in length, then see if the text will flow better if you combine some of them. The best test of your writing is to read it out loud. If you find it flows well and you can breathe in logical places, then you are probably about right.

6.3 Punctuation

Punctuation often causes students significant problems and written work may end up with commas, semi-colons, and other symbols scattered around seemingly at random. As we saw with the case of the unfortunate bar tender being shot by the panda, however, the presence or absence of a well-placed comma can completely change the sense of a sentence!

6.3.1 The full stop

The full stop is probably the easiest form of punctuation to deal with. Full stops are used to indicate the end of a sentence and are followed by a capital letter that marks the start of the first word of the next sentence. Full stops are also traditionally used to indicate the use of a common abbreviation such as 'etc.' for '*etcetera*', 'i.e.' for '*id est*' (meaning, that is), and 'e.g.' for '*exempli gratia*' (literally meaning, free example). In these cases, the full stop is not automatically followed by a capital letter. This usage of full stops for common abbreviations is decreasing and many writers simply put 'eg'.

6.3.2 The colon

The colon is most commonly used in the following ways:

- as in the previous phrase, to introduce a list or an example: 'A typical use of the colon would be as follows:...'.
- to introduce a quotation: 'In his lecture, the professor stated that: "the industrial synthesis of..."'.
- to link two contrasting statements that might otherwise be written as separate sentences: 'Homogeneous catalysis is a single phase process: this contrasts with heterogeneous catalysis...'. Here, these two statements could have been written as separate sentences, but the use of the colon emphasizes the contrast between them.
- to provide justification for a statement: 'Heterogeneous catalysis utilizes a solid catalyst: this is important when separating products from reactants'.

6.3.3 The comma

The comma is the most widespread form of punctuation and is frequently misused. Commas are used primarily as a means of providing a pause in the flow of the text; the pause can be in a variety of contexts, which are probably best illustrated by examples:

Separating items in a list

Commas can be used to separate items in a list: 'Iron, chromium, nickel, and copper are all metals that...'.

Separating out distinct phrases

Commas are used to separate out distinct phrases or clauses within a sentence. For example, we might write the sentence: 'Iron, which is a transition metal, plays a vital role in biochemistry.' In this context, the sentence would still be correct and have meaning if the clause 'which is a transition metal' was not there. However, its inclusion adds further qualification to the statement.

The presence or absence of commas in the sentence can also change the meaning of that sentence. For example, compare the following two sentences:

'The electrons which are located between the atoms form a covalent bond.'
'The electrons, which are located between the atoms, form a covalent bond.'

These two sentences have subtle but important differences in meaning. In the first sentence, the implication is that, of the electrons, only those that are located between the atoms are responsible for the covalent bond. By contrast, the second sentence implies that all the electrons are located between the atoms and form a covalent bond.

Creating pauses

Commas are used to create a pause after the opening phrase of a sentence: 'In this essay, comparisons will be drawn between…' or 'As was stated earlier, the molecular orbital model…'.

This is by no means an exhaustive list of the usage of the comma and we would encourage you to follow this up by reference to the more detailed expositions found in texts on English grammar, some examples of which are given in the further reading. As we have seen, the use of commas is important in imposing the correct structure, and therefore, meaning to the sentences we write. One effective way of checking whether you have placed your commas correctly, is to read aloud what you have written and ask yourself if the pauses indicated by the commas feel as though they are in the right place. This is by no means infallible, but it will give you a useful guide.

6.3.4 The semi-colon

The final, common form of punctuation is the semi-colon. Semi-colons are employed much less frequently than commas. The semi-colon can be thought of as being half-way between a comma and a full stop. It is typically used in the following ways:

- to separate items in a list where the sentence structure is complicated and commas are already in use to separate qualifying phrases, for example:

> 'Inorganic pigments include: titanium dioxide, which is the most widely used white pigment; carbon black, the most widely used black pigment and …'.

- to link two related sentences together rather than separate them with a full stop, for example:

> 'Titanium dioxide is a bright, white pigment; it has a high refractive index.'

6.3.5 The apostrophe

The apostrophe needs special mention in any consideration of punctuation because it is probably the most abused of any piece of punctuation although it is actually fairly straightforward to use. Increasingly, the apostrophe can be seen randomly used before or after the letter 's' when the 's' is simply indicating a plural. This is incorrect. Apostrophes have two main functions: to indicate possession and to indicate contraction.

Possession

To indicate possession the apostrophe is added to the word followed by the letter 's', thus: 'The element's valence electrons…'. The test here is whether the sentence could be re-phrased using 'of' as in: 'The valence electrons of the element …'. If the word is a plural, already ending in 's', for example 'The elements' then the apostrophe is placed after the 's': 'The elements' properties are …' (do the test: 'The properties of the elements are…'). If the word is singular but already ends in an 's', as is seen with some names such as 'James' then you still obey the rule for the singular form and add both an apostrophe and the 's', thus: 'James's lab coat was filthy…'.

Contraction

To indicate a contraction, where some letters have been omitted because two words have been joined together to make a shortened form, such as 'don't' for 'do not' (remembering, of course, the earlier point in this chapter that you should normally avoid such contractions in academic writing!).

There are, however, some traps for the unwary.

it's	who's

These both indicate contractions:' it is' and 'who is'.

Its	whose

These are both possessive pronouns: 'The method owes its effectiveness to …'. Other possessive pronouns that don't take the apostrophe are: his, hers, ours, and yours.

Try this: Break up text into sensible sentences.

Read through the following piece of text and break it into sentences using appropriate punctuation.

In the same 25 years that saw the American Revolution and the French Revolution we also saw the chemical revolution and the birth of modern chemistry the architect of that revolution was Antione Lavoisier 1743–1794 his work demolished phlogiston he classified compounds as acids bases oxides salts etc he organized the wealth of scattered chemical information into an organized system revised nomenclature invented the mass balance chemical equation and made many more contributions to chemistry he totally revolutionized the way in which chemists thought and all this packed into a relatively short 51 years of life that ended at the guillotine.

6.4 Paragraphs

Solid pages of unbroken text are very off-putting and are difficult to read, so the text is usually broken up by separating it into paragraphs. A paragraph is usually made up of several sentences that are linked together by a common theme. The break between paragraphs can then be used to separate out different sets of arguments to help structure the logical flow of the essay as a whole. For example, if you want to make four main points in your essay, then give each point its own paragraph. A possible paragraph plan might look like this:

- introduce the main idea (topic sentence);
- explain the idea (amplify the topic sentence);
- present supporting evidence or examples (quotation, study, expert opinion, or report);
- comment on the evidence (show how it relates to the main idea);
- conclude the main idea (link to the title or link to the next point).

A brief sample paragraph from an essay on the history of chemistry could be structured as follows.

Supporting evidence.

Introductory sentence, setting out the theme of the paragraph

Priestley and Lavoisier were about as different in background, ideas, and character as is possible. Priestley was born in Leeds, brought up as a strict nonconformist, trained as a minister, and worked as literary secretary to the Prime Minister. In contrast Lavoisier was a business man, geologist, astronomer, tax collector, and chemist. He inherited great wealth and married an heiress. He was a member of the French financial and government establishment. Whereas Priestley was a first rate experimentalist and simply followed where his curiosity took him, Lavoisier was a brilliant theorist who excelled in intellectual synthesis of other mens' results.

Final sentence linking back to the introductory sentence

There is no rule that prescribes how many sentences make up a paragraph but the point where you are moving from one topic to another within the essay makes a logical place to insert a break and start a new paragraph.

6.5 Abbreviations

Abbreviations are often used in scientific writing, usually to take the effort out of writing out lengthy terms, in full, each time they are used. The usual rule of thumb is that you should write out the term in full, followed by the abbreviation in brackets, the first time you use it; on subsequent occasions you can just use the abbreviation:

> 'Nuclear magnetic resonance (NMR) is a technique used to elucidate the structure of.......'

In scientific convention, some abbreviations such as 'NMR' are so commonplace that they are often used without previously being written out in full. However, the safest practice is always to write the term out in full first.

6.6 Illustrations

You should always think about using illustrations in your essays and other pieces of coursework since they can often be used to help you explain a process or describe a structure much more effectively than just by using words. Examples of the usage of illustrations could be:

- structures of molecules;
- reaction schemes;
- diagrams of apparatus;
- diagrams of atomic and molecular orbitals, bonding interactions;
- graphs and tables of results.

There are some key points to remember when including illustrations (Figure 6.1).

1. Produce your own illustrations, don't copy and paste a diagram from a website: don't forget that the rules of plagiarism apply as much to drawings as to the written word (see Chapter 10, *Avoiding plagiarism*). Also, just copying a diagram will not help you understand it.
2. Give each illustration a figure (or table) number and refer to it from the text, don't just leave it as an 'add on'.
3. Give each illustration (or table) a title so that the reader can see what is being shown.
4. Make sure that your drawings are labelled when appropriate so that the reader can identify the different structures being shown.
5. Use computer drawing packages to generate molecular structures and reaction schemes for assignments but don't forget that in examinations you will have to draw your diagrams by hand.

Remember also that illustrations can also save you words, so if you are struggling to write your essay within the word count, then a well-chosen illustration can save a lot of words in description or explanation.

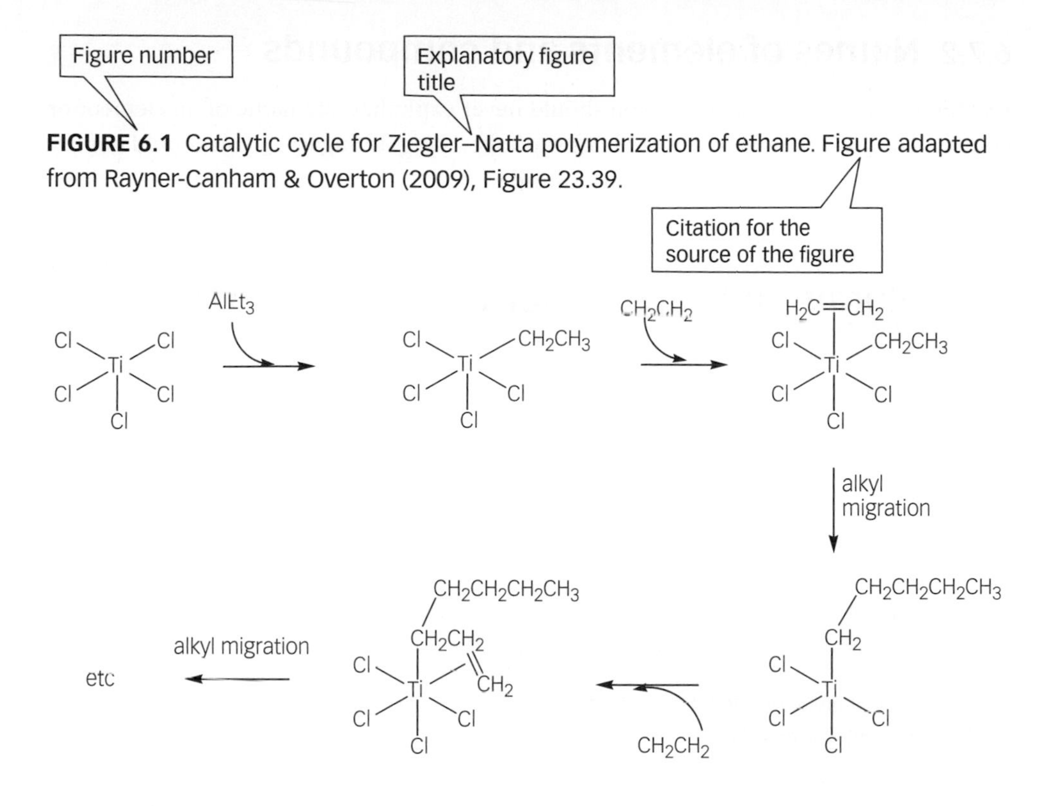

FIGURE 6.1 Catalytic cycle for Ziegler–Natta polymerization of ethane. Figure adapted from Rayner-Canham & Overton (2009), Figure 23.39.

6.7 Traps for the unwary: units, formulas, and other conventions

Chemistry uses lot of notations, abbreviations, and formulas that are not found in less technical subjects. Most of these are applied using exactly the same conventions that we use in less technical writing but students sometimes fall into the trap of treating them differently. Here are a few additional rules to be aware of when writing about chemistry.

6.7.1 Units

The proper use of units is discussed in more detail in Chapter 8, *Writing practical and project reports*. You should always use SI units unless directed otherwise by your tutor. Most units have an accepted abbreviation, for example, mol for mole, g for gram, s for second. It is important that you use the accepted abbreviation and that you are consistent. So if you use 1.0 mol dm^{-3} in one sentence do not use 1.0 M elsewhere in the same piece of writing. A very common mistake is to leave out the space between a number and its units, for example, 25cm^3 rather than the correct 25 cm^3.

6.7.2 Names of elements and compounds

Except at the start of a sentence you should never capitalize the name of an element or compound. For example, you would not say 'take 100 g of Sugar' so you should not say 'take 1 g Carbon' or '3 g of Potassium Chloride'.

6.7.3 Plurals and singular terms

We use many words specific to science that we don't necessarily come across in everyday language. Care should be taken when using such terms especially when using the plural or singular versions. So, for example, spectrum is singular, the plural is spectra not spectrums.

Try this: Plurals and singular terms

Identify the plural of the following words:
Spectrum, nucleus, formula, index, axis, analysis

Identify the singular of the following terms:
Data, phenomena, criteria

Try this: Bringing it all together

Rewrite the following (really terrible) piece of text, correcting all stylistic errors and inconsistencies:

About 2gms of V_2O_5 was put into a flask and 10 mls of H_2O, 4 ml of conc. sulfuric and l5ml ethanol were added. This mixture was refluxed for exactly 30-40 mins after which it showed a deep, blueish colouration. 5cm^3 acac was added and the solution neutralized by adding (slowly!!!) 8.8g anhydrous sod. carb. in 50 ml d.i. water.

6.8 Referencing

Finally, in this chapter on choosing the right writing style, we need to flag up referencing. We dealt with this topic extensively in Chapter 5, *Working with different information sources*, and we will return to it again in Chapter 10, *Avoiding plagiarism*. However, it is important to also note it briefly here because referencing is an important part of academic writing style. You must ensure that any information you get from another source is appropriately referenced.

* Chapter summary

Writing for a scientific audience requires a very formal style, with precise use of language. In this chapter we have looked at the conventions of writing style such as the use of the passive form and an impersonal, rather than personal expression. We have also revised some of the principles of basic grammar and construction. Finally, we have highlighted the importance of referencing, more of which can be found in Chapter 5.

You should remember to:

- construct sentences that are not overly long;
- use punctuation and grammar correctly;
- impose structure by the use of paragraphs and sections;
- use and label illustrations appropriately.

Further reading

Seely, J. 2007. *The Oxford A-Z of Grammar and Punctuation*. Oxford: Oxford University Press.

Truss, L. 2007. *Eats, Shoots and Leaves: The Zero Tolerance Approach to Punctuation*. London: Profile.

Chapter 7

Writing essays and assignments

Introduction

Essay writing is a common form of assessment in most degree courses in chemistry, both as coursework and for examinations. For many students, however, writing essays at university represents a real challenge as it is a skill that is only developed to a limited extent in pre-university courses. It may also be a long time (perhaps years) since you last wrote an essay or other piece of extended writing, and that might have been for an English course rather than a piece of scientific writing. This chapter is designed to help you appreciate what is expected of you, and guide you through the process of planning and writing essays and long written assignments. Before we begin though, consider this point: we often refer to the process of producing an essay as 'essay writing', but how much of the time is actually spent writing?

Think back to the last coursework essays you produced: how much of the total production time was spent writing and what else was involved in the process? If you are like us, then your time allocation for essays is probably something like this:

- quite a few hours searching for information in textbooks and research papers and even more time reading through the material trying to understand it!
- following that, you have to decide and note down what is specifically relevant and should be included;
- then some sort of planning stage, where you decide how to order the material and how to begin and round-off the essay,
- finally, there is the writing itself, which probably only represents about 20 per cent of the total time spent working on the essay.

7.1 What is an essay?

So exactly what is an essay? Dictionaries and websites abound with definitions of essays, viewed from different perspectives, some being more helpful than others!

In historical terms, the word essay means an attempt or trial. The *Oxford English Dictionary* includes the following definitions:

> The action or process of trying or testing.
>
> A composition of moderate length on any particular subject, or branch of a subject; originally implying want of finish, 'an irregular undigested piece' (J.), but now said of a composition more or less elaborate in style, though limited in range

YourDictionary.com defines an essay as follows:

> A short literary composition on a single subject, usually presenting the personal view of the author.

One of the problems is that the definition depends on the subject area and the expectations of the readers. Perhaps the easiest approach is to consider what we might call the 'typical' essay in chemistry. This would be a piece of structured writing, of 1000–3000 words in length, that is written as an answer to a specific question or in response to a specific title. In this context the student is expected to be able to display knowledge and understanding of the subject material, setting it out in a coherent manner and drawing logical conclusions from the material presented in order to answer the original question.

7.1.1 What are essays supposed to achieve?

This is probably a question many students have asked themselves, particularly during the early hours of the morning as a deadline is getting very close! It may also be a question that markers have asked themselves too, as they wade through a pile of exam scripts or coursework essays. Actually, a well-designed and well-answered essay question involves the development and testing of a significant number of skills, some of which are not explicitly expressed. It is all too easy to focus on the scientific content of the piece of writing and not to think much about the way that content is presented and structured, but the skill of writing is a critical element in the production of a good essay.

If we take the brief for a piece of coursework such as:

> In 1500 words, discuss the uses and applications of colloids. The essay is to be submitted to Dr Jones by 12 noon on Friday 1st March.

We can identify a list of the skills being exercised in order to complete the task as follows:

- using feedback from previously marked essays to produce a better piece of work;
- analysing the question;
- researching the subject matter;

- bringing together the information gathered from a range of sources, e.g. lecture notes, text books, research papers, web articles, etc.;
- planning the presentation of the information to address the question and make sure that all the key points are covered;
- ordering the material in a logical manner;
- framing the body of the text with an introduction, to set the scene, and a conclusion to summarize the points made;
- writing succinctly and with clarity, paying attention to the grammar, syntax, and spelling;
- referencing the information correctly to acknowledge the original sources;
- proofreading the essay to pick up on editorial mistakes and ensure clarity of writing;
- managing your time, so you are not scrabbling around trying to write the essay an hour before it is due in, even though you were given the title four weeks previously.

This is a surprisingly long list and it probably could be extended. The important point is that while some of these skills, such as researching the topic, are obvious and easily recognized as part of the process, others, such as using feedback and proofreading, are more implicit but are still very important if you are to write a good essay. As we go through the whole process of preparing and writing the essay, we will consider these different skills. Other aspects, such as researching the material, are covered in Chapter 5, writing style is covered in Chapter 6, while using feedback is discussed in Chapter 16.

7.1.2 Principles of scientific writing

Before we can get to grips with answering the essay question we need to consider some of the features of scientific writing. Most people can remember the very first essays they wrote at school (Figure 7.1), probably on the theme of 'What I did in my holidays' or something similar. These essays are usually written in an informal style and referenced very much to the individual.

Writing for a scientific audience, whether it is an essay, a report, or a research paper, is a very different exercise and requires a different, and more formal style, as we explained in Chapter 6, *Choosing the right writing style*.

FIGURE 7.1 Your first essay (probably!)

It was a lovely hot day and I went to the beach with my sister. We fought a lot and Mum got cross.

7.2 Approaches to essay writing

Producing a coursework essay involves a number of key stages, including:

- analysing the question or title;
- researching the topic;
- making a plan;
- writing a first draft;
- reviewing and re-drafting;
- proofreading and checking citations and references.

Ideally, these stages would be allocated appropriate portions of time to ensure that the production of the essay is as smooth and straightforward as possible, as represented by the timeline in Figure 7.2. The division of time won't be precise and doesn't need to be the same for each essay, but you do need to allocate time for all the important elements that make up to production of an essay, not just, for example, the writing. Importantly the timeline also includes a gap between writing a first draft and writing a second draft. This will ensure that you can be more objective in your review and make significant improvements to the work.

7.2.1 Common problems

The above represents an ideal approach and we would encourage you to aim to allocate time to your essay writing tasks in a similar fashion. If you can allocate your time in this way it is a reliable method for increasing your chances of completing an essay on time and to an appropriate standard. However, in practice the process of producing an essay is rarely as smooth and straightforward as this. Some common problems faced by students (and academics) when writing are as follows:

- leaving things to the last minute;

FIGURE 7.2 Ideal allocation of time for essay-writing tasks.

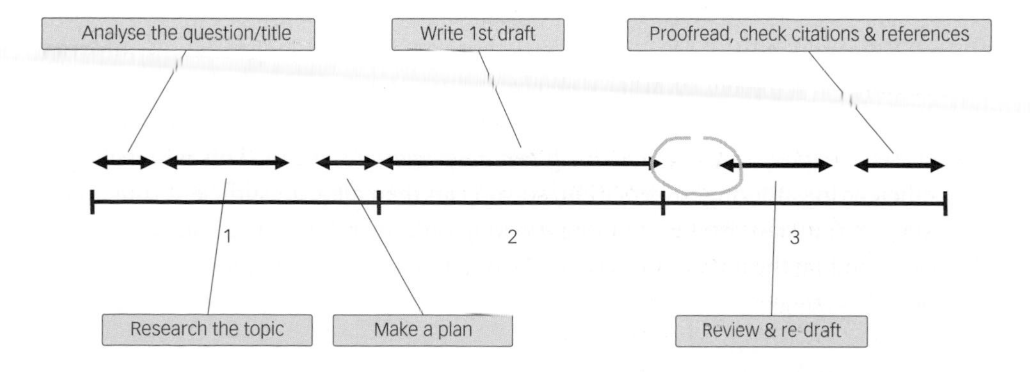

- doing too much research;
- doing little or no planning;
- taking too much time over the writing.

All these problems are real problems, many of which you may have experienced for yourself. Each of the problems has a range of possible causes, and what causes the problem will determine what you need to do to address the problem. We will deal with each in turn.

At the last minute

You might find yourself trying to complete your essay at the last minute because:

- you have lots of other deadlines to meet;
- you feel anxious about writing essays and so avoid starting them;
- you just struggle to summon up the motivation to start earlier.

If you do write an essay like this it means that you can be very focused and the work is often completed in the minimum time (because it has to be!). However, it can be a very stressful way to write and will affect the quality of your work because you run out of time to review it and so end up handing in what is effectively only a first draft. If you find yourself in this situation your immediate priority is to organize your time more effectively (as shown Figure 7.2) so that you can allocate time to improving your essay by reviewing and re-drafting.

Too much research

You might find yourself undertaking too much research because:

- you get side-tracked by information that is interesting but not relevant;
- you do a lot of reading that generates a lot of notes; and/or
- you make very detailed notes on what you read.

If you do write an essay like this it means that you have lots of material for your essay and you feel well informed, however, the volume of information can be overwhelming, not all of it is relevant and there is less time available for other stages of the essay writing process. Researching the topic is a vital task, but you don't get judged on the information you have gathered but on what you have written down. If you find yourself in this situation your immediate priority is to develop a clear focus for your reading and note making (see Chapter 5, *Working with different information sources*). This will make you more efficient in gathering information, so you can then allocate sufficient time to the other stages. You may also be spending too long on research because you are not very effective at finding the material you need. Again, have a look at Chapter 5 for some tips about search strategies.

Little or no planning

You might find yourself doing little or no planning because:

- you run out of time (see *On the last minute*);
- you find planning a bit restrictive and you would rather just start writing because writing helps you to think.

If you write an essay like this because the writing process helps you to think, it will probably mean that your writing will flow well and so it will read quite fluently. However, it can be a confusing and inefficient way to write because it is difficult to get an overview of the material and the logical sequence of the material can suffer as a result. If you find yourself in this situation your immediate priority is to set aside time for planning. The aim of planning is to provide you with an overview of the essay content so you can select and order your points in the most effective way. The essay plan can be in linear note form or it may be in the form of a visual plan (see Section 7.3.4), but whatever the format, it will provide you with instructions for selecting and ordering the content of your essay.

Writing takes too long

You might find yourself taking too long over the writing because:

- you re-write individual paragraphs again and again;
- you write too much then struggle to cut it down to the word count;
- you simply find getting the words down on paper very difficult.

If you find yourself in this situation your immediate priority is to speed up the production of the first draft in order to have more time for reviewing and re-drafting. In this case, you need to have a plan of action for the writing process. Start with a set of bullet points that summarize what you want to say, and then just write a paragraph about each point, allowing yourself an approximate number of words for each paragraph as a proportion of the essay as a whole. Don't try re-writing any sections until all the paragraphs are written. When you have the body of the essay in place, you can think about linking the paragraphs together. At this stage, you will have a good appreciation of how long the essay is going to be and also of the overall structure of the essay. Now you can devote some time to editing, or re-structuring the essay.

7.3 Producing a coursework essay or assignment

We can now use the steps identified in Section 7.2 to show you how to produce a coursework essay, based on the following stages of production:

- plan your use of time;
- analyse the question/title;
- research the topic;
- make a plan;
- write the first draft;
- review and re-draft;
- use feedback effectively.

7.3.1 Plan your use of time

You will have noticed from the common problems identified above that crucial to improving your writing is managing your time. Many of us are not very good at time management. When we are set an assignment and told that the deadline for submission is not for six weeks, the natural reaction is often to put it to one side and forget about it until a few days before the submission date. That means many of us err towards being 'on the last minute-ers'. We all know that this can lead to problems.

- 'I've only two days in which to write the essay and have just found out I can't get hold of a copy of the textbook because other students have already borrowed it.'
- 'The library doesn't have access to the journal article I really need and I have to find something else. Help!'
- 'The essay for this module is due in at the end of the week but I've also got two other deadlines coming up...'.

One thing to remember when an assignment is set is that the deadline is the latest possible date you can hand it in – not necessarily the hand-in date. Tutors won't mind if you hand in your essay before the deadline!

So when should you begin working on an essay assignment? The obvious answer is 'as soon as you are given the title and your instructions'. It may not be practicable to race off and start writing immediately, but you do need to incorporate your assignments into your course planning. Clearly, the sooner you start the better, but you may need to consider some other factors as well. For example, although you have to submit the essay in week 8 of the term, you know that you won't finish covering the topic in lectures until the end of week 4, so you are probably better off not trying to do anything until you have the background from the lectures. An example is shown in Figure 7.3.

One useful approach to your planning is to put all your assignments onto a single chart or diary. For example, you could use a calendar. When planning, it is best to work backwards from the submission date and break the assignment down into its separate elements, as described above. This also has the advantage of giving you positive feedback because you can reward yourself as you cross each section off on the calendar. If you make sure that you incorporate the submission dates for other pieces of work as well, then you can plan your time effectively and also avoid hitting an unexpected log-jam of work that needs doing and handing in all at the same time.

FIGURE 7.3 Planning your time for an essay.

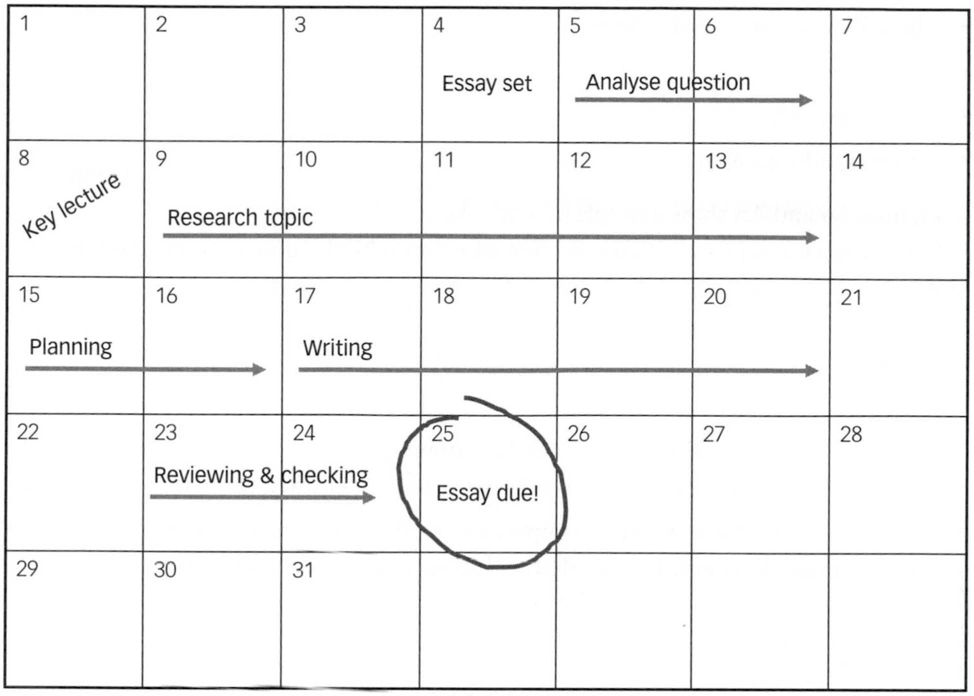

1	2	3	4 Essay set	5 Analyse question →	6	7
8 Key lecture	9 Research topic →	10	11	12	13	14
15 Planning →	16	17 Writing →	18	19	20	21
22	23 Reviewing & checking →	24	25 Essay due!	26	27	28
29	30	31				

When planning your timeline, you need to be realistic about allowing yourself enough time for the different components of the work, for example carrying out research can be very time consuming, particularly if you need to search for and read through (and understand!) a number of research papers. By the same token, you also need to be strict with yourself when undertaking the research and not get carried away looking for more and more papers. Obviously, you won't be spending all your waking hours during those two weeks working on the essay: you will have other commitments such as lectures and practical classes, as well as keeping time for socializing. Therefore within your timeline, you will need to decide on what days you are going to focus on the essay and for how long. Again, the key features are that you should try to be realistic, and then do your best to stick to your schedule.

What happens if things go wrong? Advance planning is, again the key: whenever possible, leave a gap between your planned completion of the work and the submission date, then you have some leeway if something unexpected throws your plans out of line.

7.3.2 Analyse the question or title

It is very important that you write the essay that is asked of you: there is no point in writing an essay that doesn't address the question asked, however brilliant that essay may be. So, you have your time plan in front of you, and are ready to go. STOP. Before you do anything else, read the question carefully and then read it again, highlighting the following features:

- the subject of the question/title;
- the instruction;
- the key aspect;
- other significant words.

Figures 7.4 and 7.5 show a couple of worked examples for you.

Sometimes it might seem that you are having to highlight almost every word, but it is still worth doing if it helps you register all the important pointers.

There are several standard instructions used for essay questions, some of which appear to be very similar but with some differences in emphasis. The most common instructions are:

- *Compare* – is asking you to identify and comment particularly on features in common but also aspects of difference between two or more identified elements;
- *Contrast* – this is also a form of comparison but focusing predominantly on the areas of difference. This may be included with compare as in: 'Compare and contrast…'

FIGURE 7.4 Analysing the question/title, example 1.

Discuss the uses and applications of colloids.

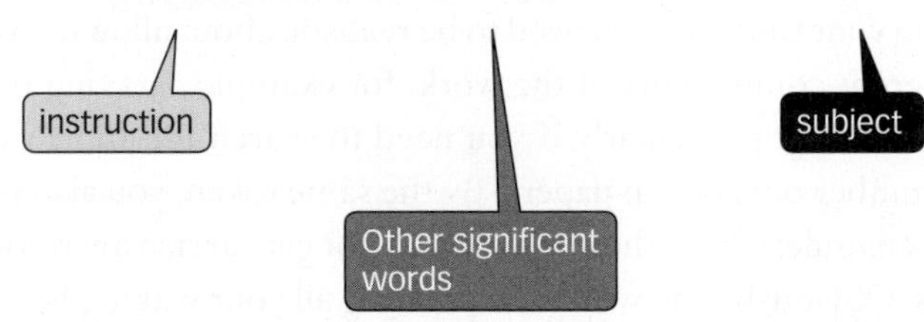

FIGURE 7.5 Analysing the question/title, example 2.

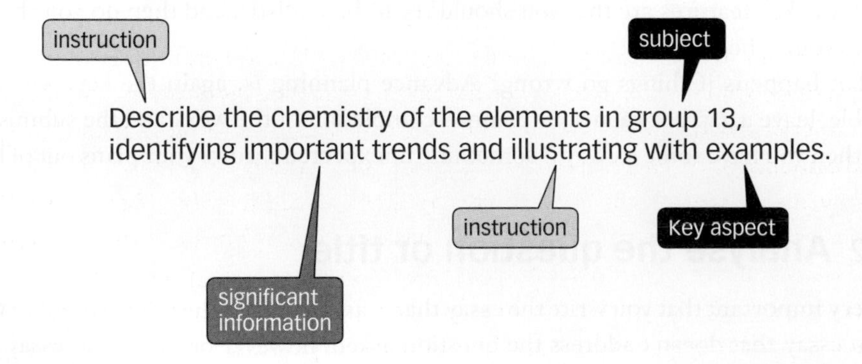

thereby emphasizing that both the similarities and the differences must be brought out with equal weighting;

- *Describe* – is asking you to write a detailed description of the subject;
- *Discuss* – is often used to mean the same as describe, but it can also be used in its more literal sense of discussing the merits or shortcomings of a particular statement or idea, as in the example given above;
- *Explain* – is often used in the context of making clear how or why something happens, for example: 'explain why heterogeneous catalysts are more environmentally friendly than homogeneous catalysts';
- *Evaluate* – the strict interpretation implies some form of quantitative assessment of information but this may commonly be used in the more qualitative context of evaluating an idea;
- *Give an account of* – this usually has the same meaning as describe;
- *Write an essay about* – again this typically means describe.

Note the example in Figure 7.5 includes the word 'and'. This was to ensure that both sections of the title were addressed: 'Discuss… and illustrate…'. Although the first section may be the more important of the two, you need to give sufficient time and effort to the second aspect, even if it does not necessarily amount to being half of the essay. The pitfall to avoid is writing a very long first section and then suddenly realizing that you still have the second aspect to deal with, so that it is tacked on, very clearly as an afterthought.

Once you have identified what basic type of essay you need to write, check for some other key pointers that give you some specific details about the approach to be adopted. These may include the following:

- *Write an illustrated account*—this is emphasizing that you need to include diagrams or other forms of illustration;
- *Using named examples*—make sure that you use specific examples that illustrate your arguments;
- *With reference to the recent experimental evidence, discuss the process…;*
- *With reference to the research literature, compare the model for… with….*

These groups of titles clearly show the emphasis that the question setter wants you to take in writing your essay. For example, it is not just a factual description that is required for the last two examples: the essay should be heavily weighted towards using the experimental evidence or recent research papers to support your arguments.

Finally, make sure that you are very clear about the breadth of the subject material required in your essay: look for the key aspect that specifies a group or range. For example, in an essay on main group chemistry, the title might refer specifically to the oxides or Group 16; a title about catalysis might refer to a specific type of catalyst; or a title about pharmaceuticals might refer to a specific group of compounds.

7.3.3 Research the topic

Researching the topic is probably the most time-consuming part of the whole exercise. Your lecturers will no doubt hope that you are really enthused by the research you undertake and that you explore the subject in depth. That's great, but you need a good dose of pragmatism in your approach and to be aware that undertaking the research could be literally an endless task. Therefore, you need to keep your timeline in mind and, whilst working as efficiently as possible, be very clear that you will need to set specific limits to the amount of research you undertake. You can also save yourself a lot of time by some initial investment: sit down with a blank sheet of paper and brainstorm some ideas – that is simply write down, in any order, all the ideas you have – about what types of information you need for the essay.

You will also need a strategy for your research: particularly when using online journals and other web-resource search engines, it is very easy to end up with very large numbers of hits, many of which are not directly relevant to your needs. See Chapter 5, *Working with different information sources*, for a discussion of approaches to searching.

Use appropriate sources

There is an increasing amount of material available for research purposes and you should be strategic and critical in the way in which you use these different resources. These resources are dealt with in detail in Chapter 5, but a summary is also given here. The main sources of information are listed below and it is probably a good strategy to work through them in this order in order to keep a clear focus. This approach also means you are starting with relatively simple, well-defined information and progressively moving to the more complex, wider sources.

- Lecture notes and handouts – these will often provide the starting point for your investigations, by reading through what your lecturers have already told you about the topic.
- Textbooks – you can use textbooks to expand your knowledge and understanding of the principles of the topic. If you are in the very early stages of your university career, textbooks may provide the bulk of the information you need for the essay.
- Reading lists – for many modules you will be given lists of suggested reading, that provide additional subject depth and breadth, to enable you to develop your understanding of topics covered in lecture courses. These lists will normally comprise journal articles and similar materials that have been selected carefully by your lecturers as being appropriate for the module you are studying. You may even be given some suggested titles as a starting point for your essay assignments.
- Research journals – as you progress through your course, there will normally be an increasing expectation that you will be reading and using research papers that you have found in research journals in order to develop the subject content of your essays. Through your library or computer service you will probably have access to useful

search engines such as Web of Science that you can use specifically for searching through the vast number of research journals that are available in electronic format. As we mentioned earlier, you will need to be careful in developing your search pattern. You can do this by selecting the appropriate key words and seeing how many hits you obtain. Gradually refine your keyword selection until you have a limited number of hits that should be closely relevant to the subject matter of your essay. You will probably still have several tens of articles in your list and reading all those in full would clearly be very time consuming. Reading the titles of the papers your search has come up with is therefore the next stage of selection. When you have drawn up a manageable short-list of the titles that appear most relevant, then you can go to the next stage of reading the abstracts and from your final list, select those papers that you need to read in full (having first checked that you can get access to the full articles).

- Web resources – there is a vast and ever-increasing volume of material available on the web and it varies in quality from the excellent to the downright wrong. When you are looking through the journals of professional scientific bodies, you have the confidence of knowing that the research papers have been reviewed by scientists who know the subject area very well. Therefore, you can be confident about the validity of the material being presented. By contrast, the material that is posted on websites may have been posted by experts in the field but it may also have been put there by people who have little or no real knowledge but who have views about a subject that they want to broadcast to the world. So, you need to be cautious and critical and look for specific signposts that can indicate the reliability of the information presented: for example, if the website is published by a university or a professional body, then the information is likely to be accurate and reliable. If it is published by a commercial organization, for example a drug company, it too is likely to be accurate but may also be subject to a certain amount of marketing 'spin', for example to encourage take-up of a specific drug. At the other end of the spectrum, if the material is published by an individual or organization with no obvious professional qualifications or academic affiliations, then you should treat the site with considerable caution. Mention here must also be made of Wikipedia. Wikipedia is an online encyclopaedia to which anyone can contribute. It is continually increasing in volume but, because of its nature, the quality of the material is variable: some is of high quality and has been written by experts in their area. Some is much less accurate and is not necessarily reliable. Some is wrong. Unless you are an expert who already knows the field, the problem you will have is that of picking out what is good from what is not. It is seldom acceptable to cite Wikipedia as one of your sources in an essay or assignment.

Make relevant notes

You now have in front of you a large amount of information that you have gathered from your research. Most of it will be in the form of articles extracted from different sources. Some of these articles may be quite lengthy and all of them will contain more material than is needed for your essay. So the next stage is to extract the relevant sections from these articles and put them in note form in your own words. The section on *Working with texts* in Chapter 5 should guide you here.

7.3.4 Make a plan

There is often a great temptation, having done your background reading, simply to sit down and start writing. This is, generally speaking, not a good idea: it is very easy to drift off the topic of the essay, to leave important points out, or to produce a very disjointed essay. Planning is a very important part of essay writing and will help you gain the best marks you can by allowing you to produce a coherent structure to your essay.

When you do the planning, however, will depend on your approach; there is more than one way of using plans in essay writing. An essay plan can be used before or after the writing of the first draft, and your preference will depend on how you approach the task of writing essays.

As discussed earlier in this chapter, conventionally, the order would be:

- analyse the question or title;
- research the topic;
- make a plan;
- write a first draft;
- review and re-draft;
- proofread and check citations and references.

But if you are one of those people who struggle to write an essay plan, what should you do? The answer to this question depends on why you struggle to plan. Consider the descriptions below and think which approach best describes your way of working.

Writer 1

I like to make plenty of notes before starting. I like to have a good idea of what I am going to write before I begin the first draft. I find it difficult to begin a draft if I am not clear on the direction my writing is going to take.

If this is you then you need to use essay plans before you begin the draft, as per the conventional approach.

Writer 2

I tend to make fewer notes than most people and tend to want to start writing as soon as possible. Often I don't know what I want to say before I start to write, but I find that ideas occur once I have started. For me, most of my thinking is done through the process of writing.

If this is you then you need to set an early deadline for the completion of the first draft so you have plenty of time to review and reorganize the content. This is important because the content of your first draft will only be organized in the order in which the ideas occurred during writing. You will therefore need to spend more time on reviewing the draft than a person who has planned the structure beforehand. Your approach to writing can be very effective, but you must allow lots of time to review and improve the structure of the first draft.

Writer 3

I like to do plenty of notes before hand and probably have a general of idea of the essay content before I begin to write. Once I have started to write I find that new ideas occur to me so the content of my draft is often different from my original intentions.

If this is you then you need to use essay plans before you begin the draft, as per the conventional approach.

Let's go back to one of the essay titles we discussed earlier:

Describe the chemistry of the elements in group 13, identifying important trends and illustrating with examples.

There are many different ways of planning your essays, so we will look at just two different formats: the linear plan and the visual plan.

The linear plan

An example of a linear plan is shown in Figure 7.6. This is very useful if you like producing lists of key headings, to which you can then add subheadings and specific details. Having drawn up the list of key points, you can easily re-order them to create a logical structure for the essay. You can also see very quickly, by referring to your notes, if you have any gaps in your research that need some additional detail before you start writing. Furthermore, once you know how many key points you have, you can divide up your word allocation between the different points, with the target of giving a paragraph over to each point. When you do start writing, you can tick the points off as you go through them, and so make sure that you don't leave anything out. This approach also means you are not likely to drift away from the topic as you can if you just start writing without any planned sequence to the essay.

Visual plan

An example of a visual plan for the same essay is illustrated in Figure 7.7. Here, the plan takes the form of linked ideas that are gradually developed into more specific details. This is often a very useful approach if the essay is conceptually difficult to put together and you are not sure how you can structure it. Once you have your basic plan drawn, you can allocate the numbers of words to the different levels so that each major branch is given appropriate weight. Again, as you write, you can tick off the branches that you travel along. For example, starting from the centre, you can tackle the physical

FIGURE 7.6 An example of a linear map for writing a course essay.

Introduction (approx 200 words)

Overview of the group, list elements, metals, metalloids, non-metals.

Body of the essay (approx 1600 words)

- Properties of the elements (300 words)
  - Physical properties
  - Occurrence, extraction, uses
- Simple compounds (800 words)
  - Need to mention electron deficiency, how is it removed
  - Hydrides
  - Oxides
  - Halides
  - Bring in trends e.g. oxidation states
- Point 3 (300 words)
  * ...
- Point 4
  * ...

properties of the elements first, followed by their occurrence, extraction, and uses. When you have completed this topic, you return to the centre point and start along the next main branch.

FIGURE 7.7 Developing a visual plan for an essay.

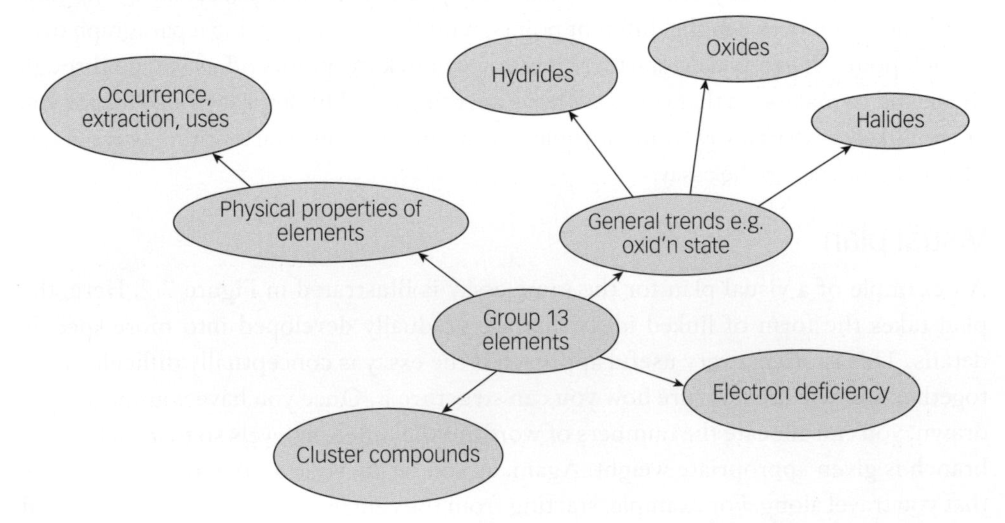

Try this: Produce an essay plan

Imagine that you are going to write an essay on your favourite hobby (so that you don't have to do any research). Produce a plan for your essay. Get someone to give you some feedback on it. Does it look logical and well structured?

7.3.5 Write a first draft

There are three main sections to an essay that you will need to write:

- the introduction;
- the body of the essay;
- the conclusion.

We will look at writing each of these sections later in the chapter but it is useful to have in mind a structure for the essay before we begin trying to write. One way of summarizing what these sections comprise is: 'tell your readers what you're going to tell them, tell them (in detail), tell them what you have told them'.

The introduction to the essay is used to frame the question and to set out what aspects are going to be addressed. An introduction is typically one or two paragraphs in length and should certainly not comprise more than 10 per cent of the essay word count. The body of the essay is the bulk of the essay. In this section you set out the arguments, in detail and in a logical order. Finally, the conclusion is used to draw the reader back to the main focus of the question and to summarize the key points that have been made in the body of the text in a way that clearly addresses the question. As with the introduction this should not normally exceed 10 per cent of the essay.

Body of the essay

Regardless of when and how you do your essay plan, you should have gathered together from your research the evidence to support your points. The inclusion of this evidence is a very important part of scientific writing. It is not sufficient to state:

Inorganic arsenic is excreted in urine.

Each statement should be supported by a specific example, so:

Inorganic arsenic is converted to methylated metabolites, and most is excreted in urine as dimethylarsinic acid in humans and animals[15].

or

Inorganic arsenic is converted to methylated metabolites, and most is excreted in urine as dimethylarsinic acid in humans and animals (Naranmandura, 2006).

Note that the statement of the example is supported by a citation, indicating where the idea you have described came from (see the section on *Citations and references* in Chapter 5, *Working with different information sources* for more information on referencing). As you write each section of the body of the essay, make sure that you mark off each point on the plan and then make sure that the paragraphs link together, rather than appearing as completely independent sets of statements.

If, for example, you want to make five main points in your essay, then give each point its own paragraph. A paragraph is usually made up of several sentences that are linked together by a common theme. The break between paragraphs can then be used to separate out different sets of arguments to help structure the logical flow of the essay as a whole. A possible plan for each paragraph might look like this:

- introduce the main idea (topic sentence);
- explain the idea (amplify the topic sentence);
- present supporting evidence or examples (quotation, study, expert opinion, or report);
- comment on the evidence (show how it relates to the main idea);
- conclude the main idea (link to the title or link to the next point).

Introduction

The introduction is a signpost for your reader, showing how you intend to answer the question. Remember what we said above about the structure of the essay: in the introduction you are going to tell the reader what the essay is going to cover. One possible structure for an introduction could be:

- begin with a general point about the central issue;
- use the words of the title to show how you will focus on the question;
- indicate what the structure of your writing will be;
- make a link to the first point.

Note that in our consideration of the writing process, we have placed the introduction after the main body of the text. There is no rule that states that you have to write the sections in the order in which they will appear in the final document. Some people find it is logical to write that way, but others prefer to write the introduction after the rest of the essay, because then they have a clear picture of what it is they are introducing.

Conclusion

The conclusion is where you summarize for the reader your main points, linking back to the question and highlighting the most important aspects of your material. It acts as another signpost to your reader, rounding off the essay. As such it gives you the opportunity to:

- refer back to the question to show your essay directly addressed the set question;

- summarize briefly for the reader the main points your essay covered;
- show the overall significance of the material in relation to the question;
- provide an overall assessment of theories or arguments, summarizing your own viewpoint.

Again, you might want to write this in a different order from the final document: for example, you might like to have a conclusion written out first because you can use it as a guideline for reminding yourself where the essay is heading.

7.3.6 **Review and re-draft**

Reviewing and re-drafting are important stages of the whole production process, but they often get overlooked because not enough time has been left to do them effectively. Again, this is where planning your use of time is so important: you should aim to leave yourself time after writing the original draft to review what you have written rather than simply aiming to get the basic writing done by the deadline. Indeed, review is most effectively done when you have left a gap of a few days between writing the draft and undertaking the review. As identified in Figure 7.2, if you read through the essay immediately after you have written it, it is often difficult to look at it critically.

There are three key stages to the process of review.

1. Reading through the essay critically, ask yourself:
    - Does the essay answer the question?
    - Is there a logical flow from one paragraph to the next?
    - Does the introduction set the context of the essay and explain what is being written about?
    - Does the conclusion summarize the key points and set out the conclusions you have drawn?
    - Have I put in citations to support the key statements in the text?
    - Are any diagrams or illustrations in the right place and have I included the source reference if necessary?
2. Re-drafting the essay to improve the structure following your review.
3. Proofreading the essay to:
    - correct errors of spelling and grammar;
    - check the citations match the reference list.

Re-drafting

Assuming that you have written your first draft in good time, you may wish to re-draft the essay. The extent to which writers may wish to write more than one draft of an essay depends on how careful the initial writing stage was (as illustrated with the examples of writers 1, 2, and 3). Some writers prefer to get something down on paper (or computer) and then juggle around with it until they are happy with the structure; others work very closely to a structured plan, and therefore may only write a single version of the essay.

Re-drafting can take two stages: re-ordering of paragraphs to help improve the flow of the essay and then re-writing individual paragraphs to improve the way in which the points are conveyed to the reader. If you have written the essay using a word-processor, then re-ordering is relatively straightforward by cutting and pasting the paragraphs into a different order. Likewise, with the individual paragraphs, it is easy to re-write a paragraph and insert it back into the text.

Don't forget to re-read the whole essay, to make sure that your re-ordering of the paragraphs hasn't broken the links between the different points.

Proofreading

It is very tempting, when you have finished writing the last word of the essay, just to print it out and sit back feeling satisfied that you have finished and it is on time (maybe only just!). It is very important, however, that you proofread the essay, checking for errors in your typing, spelling, and grammar. There is a danger that writers become too reliant on the spelling and grammar checkers in word-processing packages and assume that because the program has not highlighted a piece of text, then it is correct. These packages are not infallible. For example, many typographical errors will in fact result in a valid word so the spell-checker won't highlight it. For example 'to' and 'too' are both valid words but they clearly have different meanings. Imagine if you misspelt 'lattice' as 'lettuce'. Conversely, many programs will flag up many of the technical terms you may use in chemistry. For example, the word-processing program used for this book identified 'methylated' as being incorrectly spelt. As a consequence your essay on screen may be highlighted in lots of places and so there is a tendency to ignore all of these as 'false positives'.

How should you go about proofreading? One of the hardest parts of proofreading is that, when we read through something we have written, we read what we expect to see and can very easily overlook mistakes. This is especially true if you have just finished writing the essay and are still feeling totally immersed in the exercise. If you have managed your time carefully you will have at least a few days spare before you have to hand the essay in. So, put the essay away for a few days and then sit down and read it carefully, as if it had been written by someone else and you were tasked with finding errors.

If you look to the punctuation section in Chapter 6, *Choosing the right writing style*, the best way of checking this is to read the essay out loud, as if to an audience. You might feel really stupid doing it, but you can always shut yourself away so no-one will know! As you read make sure that there are breaks in the text (and that you don't run out of breath!) and that they come in the right places for the sense of the text. Reading out loud is also a good test to make sure that the structure as a whole makes sense.

Another useful exercise is to put a guide (e.g. a ruler) underneath each line as you read through, so that you focus on the words you are reading. Doing that, you are much more likely to pick up on typographical errors that you would not spot otherwise (and that your spell-checker won't pick up either), such as 'to' instead of 'too', or 'lose' rather than 'loose'.

Finally, check your use of references:

- make sure that your key statements are supported by citations in the text;
- check that each citation in the text is matched against a reference in the listing at the end of the essay;
- likewise, check that each reference at the end appears as a citation in the text.

Is proofreading really that important? Yes, it is for a number of reasons:

- the person who is marking your essay will not be impressed by reading text that is littered with errors and, as a consequence, you will probably get lower marks than the scientific content alone might merit;
- some mark schemes include a specific allocation of marks for grammar and spelling;
- errors in punctuation and syntax can lead to ambiguities of meaning, so the sense of what you are writing is no longer clear;
- it is good to get into the habit of writing accurately: if you are asked by an employer to write a technical report, that employer will not be pleased if there are numerous mistakes;
- furthermore, when you are trying to sell yourself in job applications, these mistakes can be very costly: employers have been known to say that the first filter they use when processing applications is that any application with three or more spelling mistakes is automatically rejected!

So, get into good habits early and always check what you have written.

7.3.7 Use feedback effectively

It is very tempting, when receiving a piece of marked work, to look at the mark or marks obtained, feel moderately satisfied (or disappointed), and then put the essay away and not look at it again. It may well be that, since you submitted the essay, you have moved onto another module and don't feel that any comments would be relevant to what you are doing now. If you do this, you cannot hope to improve the way you write because you are not paying attention to the guidance given as to where you went wrong or how you could have done better. Likewise, it is also important to know what things you are doing particularly well, so you can continue doing them.

Using feedback is a very important part of the learning process and we go into more detail on this topic in Chapter 16. Feedback on essays should come in two forms: there should be subject-specific feedback, for example the identification of factual errors or omissions, and more generic feedback related to your writing style. The subject-specific aspects should help improve your understanding of the topic and may come in particularly useful when you come to revise in preparation for examinations. The generic feedback should enable you to improve your essay writing for the future, irrespective of the topic. This generic feedback might address aspects such as the use of references, the layout of the essay, or the use of illustrations.

So, before you consign your essay to the back of your cupboard, take time to read through it carefully: re-reading an essay after a few weeks is itself a good exercise because you will have a very different perspective compared with when you were immersed in writing it. Read through and think about the feedback you have been given. First, make sure that you understand why the essay as a whole got such good or bad marks: was it focused on the question? Did you go into enough detail? Did the essay have a clear structure with each point being explained and then leading on to the next? If you don't understand where you went wrong, make an appointment to see your lecturer so you can talk through the aspects you don't understand.

Reflecting and acting upon feedback is a key aspect of effective personal development planning (see Chapter 17). You should be able to learn from the feedback and plan for improvement. Your essay and the feedback on it can be logged as evidence of your written communication skills. Keep a record of the key points of feedback from your different pieces of work, note down the generic feedback from the current essay, and then look back over the previous pieces of work to check whether this is an issue that has arisen before. If it is a problem that has been flagged up more than once before you might well consider seeking specific advice about how to tackle it. Having noted down the key points, write a list of a few action points to bear in mind for the next essay you write, such as:

- write a more detailed plan;
- write shorter paragraphs;
- check the referencing style and that all references cited in the text are listed in the references.

Then, when you come to write your next essay, refer to your list of action points and make sure that you improve on those aspects. In this way, you should be able to maximize the benefit from the feedback given and progressively improve your writing. Don't forget, though, if you can't read the feedback comments or don't understand them, don't just give up: go and ask your lecturer for help.

7.4 Writing essays in examinations

Most chemical science courses at university will require you to write essays under examination conditions at some stage. In this section we are focusing on some tips in relation to essays; further guidance relating to examinations in general is given in Chapter 15,

Getting the most out of exams. So, you are sitting in the examination hall with the paper in front of you giving a list of titles. You don't have six weeks to think about, research and write each essay: you now only have a limited amount of time and have to write several essays in that time. After the initial panic, try to be very methodical in your approach and follow these key points: five minutes of care at the beginning of the exam can save you a lot of heart-ache later.

- Read the instructions on the paper through carefully and slowly, paying special heed to any that specify how many essays you should answer and whether you have to select titles from specific sections of the paper.
- Read the titles through carefully.
- Eliminate any titles about which you know nothing (hopefully not too many!).
- Select a set of titles on topics that you can answer, then narrow that down to the requisite number by checking the titles again, highlighting the key words, and then selecting those titles that you are confident you can address in full.
- Plan your first essay: again, time spent planning the essay and checking against the keywords in the title will help make sure that you answer the question.
- As you write the essay refer back to your plan and tick off the key points as you address them.
- Don't forget to include illustrations if they are relevant.
- Be strict with yourself about timing. It is very easy to select the first essay title as one you know most about and get carried away writing it, so that you don't leave yourself enough time to do justice to the other essays. So don't allow yourself to spend a long time on the first essay, divide your time up equally.
- Leave enough time at the end for proofreading.

If you only answer two essays instead of three, or select your titles from the wrong section of the paper, or don't address the question, you cannot hope to get good marks!

A final tip for this section: for your coursework essays you will probably have had a long time to write them, you will have written them using a computer, and probably did not write them at one sitting. Examination essays are very different: at the moment, most examination essays are written long-hand, using a pen. So, it is a very good idea to get some practice in writing under these conditions; this can form a valuable part of your revision. When you have finished revising a specific topic, look out some past examination papers and find a question on that topic. Then sit down and write an essay, under examination conditions. This will be a good test of how much you know and understand about the topic but it will also give you a feel for how much you can write in the time available. As a final guide, ask a lecturer to read through the essay and comment on it, so you can confirm that you are on the right track.

* Chapter summary

Essay writing is an important skill and as with all skills, it develops with practice. In this chapter we have shown the importance of:

- planning your use of time as soon as you have the essay title so you don't end up rushing at the last minute;
- analysing the question or title carefully so you make sure you write the essay your tutors want;
- researching the topic carefully, using the different sources available to you;
- planning the structure and order of the essay;
- reviewing, re-drafting, and proofreading to make sure you have answered the question, that the structure is logical, that the citations and references are correct and that there are no spelling or grammatical mistake.

Chapter 8

Writing practical and project reports

Introduction

The abilities to design experiments, to undertake them methodically, to observe and record the outcomes accurately, and to analyse and interpret the findings are vital to the development of a scientist. These skills, however, are not sufficient on their own: you also need to be able to communicate your findings to other people. If someone asks you for directions to find somewhere, knowing where to go is only part of the issue, you also need to be able to communicate that knowledge. In the scientific community this is done in a number of ways, the most common being presentations at scientific meetings and published reports.

If you are planning to follow a career in chemistry, you will need to develop all of these skills. However, such skills are not just of value in a scientific career: there are many careers where the skills of accurate recording, analysis, interpretation, and presentation are highly valued by employers. The aims of this chapter are to help you prepare high-quality reports. Although we will discuss fundamental aspects of data presentation, this will be fairly limited since there are other guides that go into this in great detail. Likewise, we will not discuss the processes of experimental design or data analysis. Good guidance for these aspects can be found in other books such as *How to Do Your Student Project in Chemistry* by Jardine (1994).

This chapter is laid out so that you can read it straight through, or dip into sections for guidance on how to tackle specific aspects of your work. During your studies you will be expected to read scientific papers and it is always a good exercise not just to focus on the scientific content but also to evaluate the way in which the science is presented. As you read on, it will be helpful to have some research papers to hand so you can look at the different styles of presentation.

Although the focus of the chapter is on laboratory-based practical classes, much of the material is also directly relevant to project work as well. To help create some context, the text uses as most of its examples the experiment that was used in Chapter 4, an undergraduate investigation of the hardness of tap water by analysing the concentrations of calcium and magnesium ions by a complexometric titration.

8.1 Writing your report

During your degree programme you will probably be asked to write up your practical work in a range of different formats. These may range from simply filling in tables and writing short answers to questions listed on the practical schedule you have been given, to preparing a full report in the form of a scientific paper. In this section we will take the scientific paper as the model, as it covers the most common aspects of report writing. Make sure you are clear what format of report is expected from you before you write!

8.1.1 Use an appropriate writing style

The principles of writing for a practical report are the same as those for other pieces of scientific writing and have been discussed in detail in Chapter 6, *Choosing the right writing style*. The key point to remember is that you are writing for a professional readership and therefore you need to write in a formal, scientific style.

Avoid the use of 'I' or 'We'

So the phrase: 'I carried out three titrations…….' becomes: 'Three titrations were carried out…..'. Likewise: 'On the basis of these findings, we concluded that…' becomes: 'On the basis of these findings, it was concluded that…'. This is known as writing in the **passive voice**.

Write in the past tense

You should describe what *was* done, or what *was* concluded. The only exceptions to this are when you are discussing illustrations or data presented in the report and the conclusions you can draw from them. For example: 'Table 1 shows the titres for the three sample…' or 'The conclusion of this report is that the hardness of tap water……'.

Reference your sources

Wherever you take other people's ideas, you must reference them correctly in the text and list the sources in the reference list at the end of the report. Remember that not to do so is plagiarism (Chapter 10, *Avoiding plagiarism*). Referencing is dealt with in detail in Chapter 5, *Working with different information sources.*

Use SI units

Measurements should normally be presented using the SI (Système International d'Unités) conventional form. See Box 8.1 for details of some common units.

BOX 8.1 Common units and the SI system.

The SI system of units is based on seven 'base units' that are the fundamental measurements from which the other units we use are derived (Table 8.1)

TABLE 8.1 The seven base units of the Système Internationale d'Unités (SI units).

Quantity	Unit	Symbol
Length	metre	m
Mass	kilogram	kg
Time	second	s
Temperature	kelvin	K
Amount	mole	mol
Current	ampere	A
Luminosity	candela	cd

From these base units we derive a series of units for different measures. Table 8.2 lists some the most common units that you may come across in practical work and their definitions.

TABLE 8.2 Common units and their definitions.

Measure	Unit	Symbol	Derivation
Area	square metre	m^2	
Volume	cubic metre	m^3	0.001 m^3 1 dm^3 (1 cubic decimetre, the SI unit for volume)
Volume	litre	l	
Velocity	metre per second	$m\ s^{-1}$	
Acceleration	metre per second per second	$m\ s^{-2}$	
Force	Newton	N	$kg\ m\ s^{-2}$
Energy, work	Joule	J	N m $kg\ m^2\ s^{-2}$
Power	watt	W	$J\ s^{-1}$ $kg\ m^2\ s^{-3}$

BOX 8.1 Cont'd

TABLE 8.2 Cont'd

Measure	Unit	Symbol	Derivation
Concentration	mole per cubic decimetre	mol dm^{-3}	
Temperature	degrees Celsius	°C	K
Electric potential difference	Volt	V	W A^{-1}
Electric resistance	Ohm	Ω	V A^{-1}

Note, when writing compound expressions, e.g. m s^{-1} you should use a space to separate the terms. This avoids confusion where terms could be misinterpreted, otherwise, for example N m could be mistaken for nm (nanometre) when hand-written.

8.2 Get the structure right

The standard format of a scientific paper is as follows:

- title;
- abstract;
- introduction;
- methods;
- results;
- discussion;
- references.

You may also wish to include some additional sections such as Acknowledgements, Appendices (See Results), and Keywords or a list of abbreviations used. The same structure can be used for a practical report.

We will look at each of these sections in turn and you may find it useful to refer to some of the papers from your reading lists to look at the way these sections are organized and presented.

8.2.1 Title

The title must be brief and informative, so that the reader can see at once what the paper is about. In the case of scientific papers for publication, the title is very important as it will be used as part of the information for electronic searching, so authors will often include keywords to increase the probability of the paper being picked up by an electronic search.

Often journals will limit the number of characters that can be used in the title. A typical limit would be 120 characters, including spaces, so it is good practice to write your titles within this limit.

Have a look at the four sample titles below, all of which describe the same investigation.

1. An investigation into the temporary and permanent hardness of local tap water by a titrimetric method utilizing the sodium salt of ethylenediaminetetraacetic acid.
2. Investigating the hardness of tap water by titration with EDTA.
3. How hard is tap water?

Which title provides the key information most efficiently?

The first title is unnecessarily long, though it does contain all the essential information. The third title is too brief since it does not give any indication of the basis of the experiment and is too informal. The second title contains all the essential detail for the reader to know whether to read further or not.

8.2.3 Abstract

There is a real art to writing a good abstract. The abstract must provide the reader with all the essential information from the paper within a very limited number of words, typical abstracts being around 150–200 words in length. In many cases, the abstract may be used as a free-standing source of information: for example when you are researching for an essay, you may well only read the abstract of a paper rather than the whole paper.

The abstract will normally be the last part of the paper that you write and it is worth going through the following checklist to make sure you have covered everything.

- Background: very brief but enough to set the context for the…;
- Aims: the key question you set out to answer using the…;
- Methods: again these should be described very briefly, summarizing the technique(s) used to obtain the…;
- Results: here you need to quote the key findings of the investigation, including the actual values obtained from which you drew the…;
- Conclusions: the key points you draw out from the results. The conclusions should be related back to the initial aims.

The results and conclusions are the most important part of the abstract because, in a scientific paper, this is where you are stating what you have found that is new. These two aspects should therefore make up the bulk of the abstract, with the background, aims, and methods being written as succinctly as possible. When you read through the abstract of your report (or better still ask a friend to read it through), the two points that should be immediately obvious are: what was found and why it is significant.

Compare the information content of the two sample extracts from abstracts describing our titration and identify the key pieces of information that are missing from the second extract.

Abstract 1

The aim of the experiment was to determine the hardness of a sample of tap water. This was achieved by measuring the concentration of magnesium and calcium ions in the sample. The total concentration of calcium and magnesium ions was determined by titration with EDTA solution and Erio-T indicator. The concentration of magnesium ions only was determined by titration with EDTA and murexide indicator. The concentration of calcium ions was determined by difference. The concentration of Ca^{2+} was found to be 25 ppm and Mg^{2+} 109 ppm.

Abstract 2

Tap water samples were titrated with EDTA and either Erio-T or murexide indicator in order to determine concentrations of calcium and magnesium ion. The results were found to be Ca^{2+} 25 ppm and Mg^{2+} 109 ppm.

Although the second abstract does describe the conclusion of the experiment, it does not give any indication of the initial aims. Nor does it provide any of the key information to allow the reader to appreciate what was done.

As with the title, the abstract of a paper is commonly used as a means of electronic searching and so should contain keywords that identify the nature of the study and its conclusions.

It is not normal practice to include references within the abstract.

> **Try this: Write an abstract**
>
> Go to the Wikipedia pages for either fullerene or superconductivity (or another chemical topic that interests you) and write an informative abstract that would help future visitors to the page.

8.2.3 Introduction

The introduction is the section of the report in which you describe the background and aims of the study. You might like to think of this section as addressing four questions.

- What has been done before?
- What still needs to be resolved?
- Why is it important to resolve this question?
- What do I aim to do?

A useful approach to drafting the Introduction is to start by setting out the aims of your study ('what do I hope to find out?') as a series of bullet points. Keep this list of aims in front of you while you are researching and writing the background to the study so that you keep clearly focused on the topic and don't start writing about sideline issues.

For a report in the form of a scientific paper you will need to read published papers on the topic of your investigation so that you can write a brief overview of the research that is underpinning your study. You are most likely to have to do this when writing up a practical project rather than for a routine laboratory experiment. The skill in drafting this part of the introduction lies in writing a succinct summary of the findings of the previous authors and of the experimental evidence supporting those findings. As with all such writing, it is important that you present the material in your own words and fully reference the statements (see the section on *Citations and references* in Chapter 5, *Working with different information sources* p. 64).

The background text is normally presented as a factual account without discussion of the ideas presented in those papers. In most areas of research, there are areas of knowledge that are accepted as agreed and that you can describe as the background to the study. Developing from these, there will be areas where there is still disagreement, or questions remaining to be answered. In the case of disagreements in the published research, you will need to present both sides of the argument. This is best done simply by presenting your synopsis of the two views in sequence. Thus, you might write:

Derby & Joan [2] concluded, on the basis of laboratory observations, that the main product from the synthesis was the para isomer. However, in a more recent study, Bonnie & Clyde [3] reported that meta isomer was the major product.

In the case of addressing questions still to be answered, you might phrase your statements as:

Derby & Joan [2] concluded, on the basis of laboratory observations, that the para isomer was the favoured product of the reaction. However, it is unclear whether this isomer remains the major product when the reaction is carried out under acidic conditions.

Your description of the background to the study should, therefore, lead the reader to understand the scientific basis of the study and from there be provided with an explanation of what is still not understood and why it is important that it should be understood. On that basis, you can then explain the aims of the study you have undertaken

It is inevitable that, for many of your practical classes, you will not be aiming to resolve current controversies or discover the solution to a specific question in science, though it is perfectly possible that you may be doing that in your final year research project. For the most part, your practical work will be undertaking experiments that have well-known results. Despite this, it is important that you can research the background to the study and present it as a synopsis, in your own words, of the scientific knowledge and can set out the aims of your experiment, based on that knowledge.

8.2.4 Methods

The methods section of the report is where you describe:

- what you did;
- how you did it;
- what you used to do it.

The methods section should be written in such a way that another scientist could repeat your work on the basis of your descriptions. Particularly when writing the methods section for your final year project, you may need to include more detail than would normally be found in scientific papers, for example regarding the composition of solutions used.

Think of this section in terms of preparing a recipe for baking a cake. When you read a recipe, you need to know:

- what ingredients to use;
- how much of each;
- how to mix them together;
- what type of baking tin to use;
- what oven temperature to bake it at and for how long;
- how to analyse the results (eat it!).

Although you are describing what you did, you should still remember to write in the passive voice: 'The concentration was determined by…' rather than 'I measured the concentration by …'

Check how much you need to write

In terms of reporting on class practical exercises, the necessity for a methods section is variable. For example, where you have been given a detailed schedule to follow, you may be told that you do not need to copy the methods out again. Even in this case, though, you will be expected to report on any changes made to the protocols, particularly since these may affect the results you obtain.

Assuming that you will be writing a full report, then you will need to describe the methods you used in detail. If you are using a new technique, or have been instructed to give full details of the protocols then you should set them out in full. In most papers, however, if you are using a standard procedure that has been described in detail in another paper, then you can simply refer to the previous paper and only specify any changes you made:

The calcium concentration was determined using the method of Peters [16] following pre-concentration of the sample.

Chemicals

When listing chemicals or other substances that were used you should give the name of the chemical in full (e.g. sodium hydrogencarbonate) and/or its chemical abbreviation (e.g. $NaHCO_3$), along with the amount used. If you have used specialist reagents you may also need to note additional details such as the name of the supplier. You should also note that there should always be a space between an number and its units. You would not write '100 g of Flour' so do not write '100 g of Sodium Chloride'. Be aware that names of chemicals are never capitalized.

Specialist equipment

When describing the equipment used, the amount of detail you need to give again depends on the degree of specialization. For example, if you are stating that you used 3 g of sodium bicarbonate, you don't need to give details of the balance used to weigh out the chemical. However, you might have used a specialist piece of equipment for your experiment, for example, a NMR spectrometer. In this case, you should give the details of the equipment such as the manufacturer and the model number. In some cases, it may also be helpful to include a simple diagram of the apparatus used, to show how different pieces of equipment were connected together, especially when using hyphenated techniques such as ICP-MS or GC-MS.

Statistics

We are not going into any detail of statistical analyses as there are many books for chemistry that address this topic. However, there are some key points you must remember.

For many physical or analytical chemistry experiments you will be undertaking several measurements. You may therefore need to give descriptive statistics in your methods. The basic statistics would be measures such as **percentage error,** the **mean**, the number of measurements taken (n), and the range of the values. For example, if you quote a mean value for a result, you may need also to give the number of tests carried out (e.g. $n = 27$) and an indication of the spread of the data. The spread of the data is often given in the form of the **standard deviation**, which gives the reader a measure of the confidence that can be placed in the mean value. In the methods, you would state:

Means are given with the standard deviation … the concentration of calcium ions was 109 ± 5 ppm ($n = 5$).

You may also be employing statistical tests in your analysis of the data, for example to distinguish whether a treatment had an effect on the sample population. For most practicals, you are likely to employ routine tests, such as the Student's t test. In such cases it is sufficient to state the name of the test and the level of significance that you are accepting, e.g. $p < 0.05$.

After all that

When you have completed writing the Methods section, read through it and check that you could repeat the experiments from the description you have written (think back to the recipe – could you bake the cake?).

Try this: Constructing a method

Convert this list of steps in a practical experiment into a paragraph suitable for inclusion in the Method section of a practical report.

Recrystallize product from glacial acetic acid, filter off, wash with a little fresh, cold solvent, dry in an oven at 60 °C.

Remove manganese dioxide by filtration, press filter cake well, wash with 25 ml of distilled water.

Arrange a 100 ml round-bottomed flask fitted with a reflux condenser in a beaker of water on a magnetic stirrer-hotplate.

Stir in the bath of boiling water until only a trace of permanganate remains (spot a drop of reaction mixture onto filter paper to check the colour).

Add 60 ml of water, 1.5 g of 4-nitrotoluene, 3.8 g of potassium permanganate, a pellet of sodium hydroxide, and a magnetic stirrer bar.

Filter off product, wash with cold water until free from HCl.

Cool filtrate, refilter if not absolutely clear, and acidify with concentrated HCl, checking pH with Congo Red paper.

8.2.5 Results

The results section is the core of the report. Here, you *describe* and *show* what was found along with the analyses of those findings. You should aim to lead the reader through the findings, highlighting the important features. You do not, however, attempt to interpret or explain the results in any way: this section is a factual description.

When writing the results section, you should use your data to tell a story. It is very important that the text you write can stand on its own in terms of communicating the key points. The graphs, tables, and other images are then used to support the description. A common error in students' results sections is for the data simply to be presented as the graphs and tables with little or no explanatory text, leaving the reader with the task of trying to work out what is important. So rather than write…

The results for the analyses are shown in Table 1.

It is much better to write…

From Table 1 it can be seen that there was no significant difference between the hardness of water taken from different points in the city.

In the case of the first extract, it is left up to the reader to draw their conclusions from the data shown in the table. By comparison, extract 2 provides the key summary of the observation, using the data in the table to illustrate the point.

Calculations and levels of accuracy

When using calculators and spreadsheets, be aware of the levels of accuracy of your measurements: often, a calculation performed on a calculator can result in a number with a string of up to 10 digits. For example, when carrying out a titration for our analysis of calcium ions we might only be able to measure the volume to the nearest 0.5 cm^3. So an individual measurement might be recorded as 23.05 cm^3. However, the mean calculated for all the titres, when using a calculator, might give a figure of 23.942857142 cm^3. If you put this number into your results, it implies that you could record the titre down to a precision that is not possible. As a very simple rule, don't present calculated values to any more significant figures than the original measurements.

What is the best way of presenting the results?

This is often a question students find hard to answer. There is a strong temptation to adopt the 'shot-gun' approach: if you include absolutely everything then you can't have forgotten anything! However, for the reader this can make interpretation difficult and someone marking your report may conclude that the reason for putting everything in was because you weren't sure what was important and what wasn't.

Remember that you are telling a story and so try to identify what the key points are, in sequence. Where possible you should present your results in summary form, rather than long lists of raw measurements (which can be included as an appendix at the end if necessary). If you have taken 20 measurements for a single data point, it may be more appropriate to present the data in summary form as the mean, the standard deviation and the number of measurements, rather than as a table of 20 individual measurements.

The mean reading was 187 ± 20 mV ($n = 20$).

This statement provides the reader with as much effective information as does the full list of individual measurements and is in a much more digestible form. For a practical report, you may wish to include all the raw data, in which case you can add it as an appendix to the end of the report.

The three main ways of presenting data are as follows:

- text;
- tables;
- graphs.

Text

Use statements of specific values in the text when you are referring to only one or two items, so the example used above stating the mean jump distance is appropriate for presenting in text.

Tables

Tables are very useful for presenting data in an organized form, particularly where you wish to present several variables together to give an overview of a set of results (Table 8.3). The layout of the table is important to ensure that the contents can be easily read and understood. In particular:

- try to avoid having too many columns, having more than five or six columns can make comparison of the data items difficult and also means that the text may be compressed to fit on the page;
- make sure the rows and columns have clear headings, with units, so that the nature of the data is immediately obvious;
- spread the data out and have clear delineation between rows and columns.

Give your table a number so you can refer to it from the text.

The title should allow the reader to understand the table on its own

Each column must have a heading to identify the data set and the units of measurement

TABLE 8.3 Results of titration for calcium and magnesium ions.

Titre	Ca^{2+} /cm^3	Ca^{2+} + Mg^{2+} only/cm^3	Mg^{2+} only/cm^3
1	5.30	21.45	16.15
2	5.55	21.85	16.30
3	5.65	21.90	16.25
4	5.70	22.05	16.35
5	5.65	22.05	16.40
Mean	5.57	21.86	16.29
Standard deviation	0.14	0.22	0.18

Graphs

There are many different ways of presenting data in graphical form. The types of graph you are most likely to use are:

- bar chart;
- histogram;
- scatter graph;
- line graph.

Bar charts can be used to display data related to frequency. So, if you wanted to display the actual concentration at each region numbers in each group, you would use a bar chart as shown in Figure 8.1.

FIGURE 8.1 A bar chart showing the variation in hardness of water by geographical area.

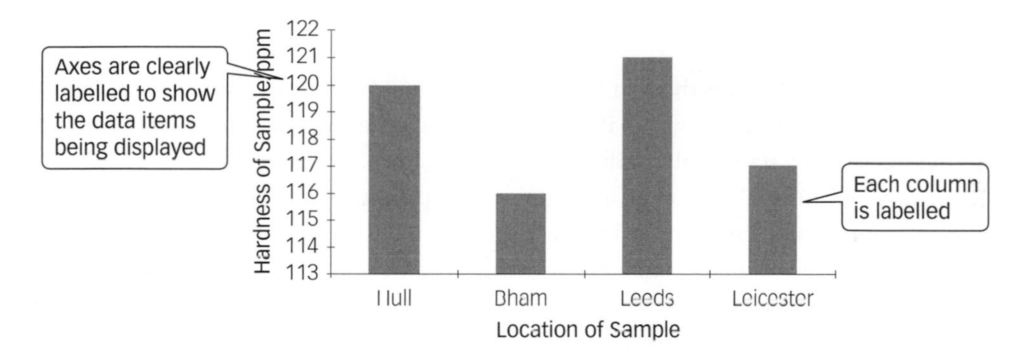

Note in Figure 8.1 that each axis is labelled to identify the information being displayed and that the scale of the *y*-axis (vertical) is set appropriately to display the spread of the data (have a look at the boxes at the end of this chapter for some examples of how not to draw graphs). A key feature of the bar chart is that the bars are separated by a space. This reflects the fact that the *x*-axis (horizontal) is displaying discrete items and not parts of a continuous population. The horizontal axis is just there as a platform for the bars.

Scatter graphs are used to display the spread of data for two variables that are related to each other. For example, the relationship between the determination of concentration of calcium by titration and atomic absorption spectroscopy, as shown in Figure 8.2.

When displaying data in Figure 8.2, there is a further decision to make, which is whether to draw a line to indicate a trend in the data. Discussion of the processes of line fitting is beyond the scope of this book: you should refer to a suitable text on statistics for analytical chemistry or chemometrics. As a rule of thumb, however, be wary of the complex line-fitting programmes that can be found in many spreadsheet and statistical software packages. These programmes can fit scattered data with very complex mathematical equations, which can be very difficult to interpret. If you think there may be a simple trend such as a linear relationship, then use the program to calculate the linear

FIGURE 8.2 A scatter graph showing the concentration of calcium determined by titration with EDTA and by atomic absorption spectroscopy.

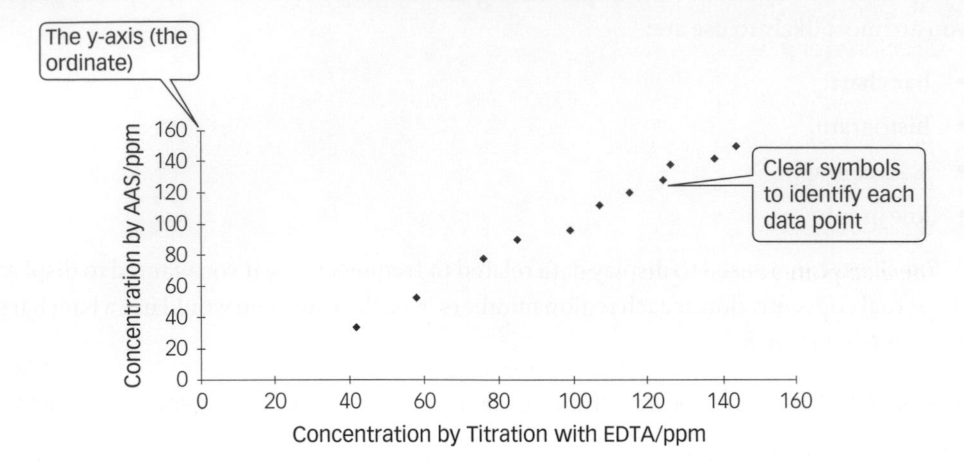

regression line that best fits the data. However, you should only draw conclusions regarding the linearity of the relationship if you also calculate the coefficient of correlation and this demonstrates a statistical significance.

Line graphs. Rather than having a scatter of data points, as for the population above, your experiment may have involved taking a series of measurements against a set of standards. An example of this would be measuring a calibration curve for UV-visible spectroscopy. How do you determine which variable goes on the *x*-axis (the **abscissa**) and which on the *y*-axis (the **ordinate**)? The variable placed on the *x*-axis should be the **independent** variable. This means it is the variable that is already known, or can be controlled by the experimenter. In the UV-vis experiment the concentration of the standards is the independent variable and the absorbance is the **dependent** variable, which is placed on the *y*-axis. The dependent variable is, therefore the variable that may change as a function of the independent variable. An example is shown in Figure 8.3.

As for the scatter plot, you will need to decide whether to draw a line to illustrate a trend in the data. This can be done either by simply joining the individual points together,

FIGURE 8.3 A line graph illustrating the relationship between concentration and absorbance for a set of standards.

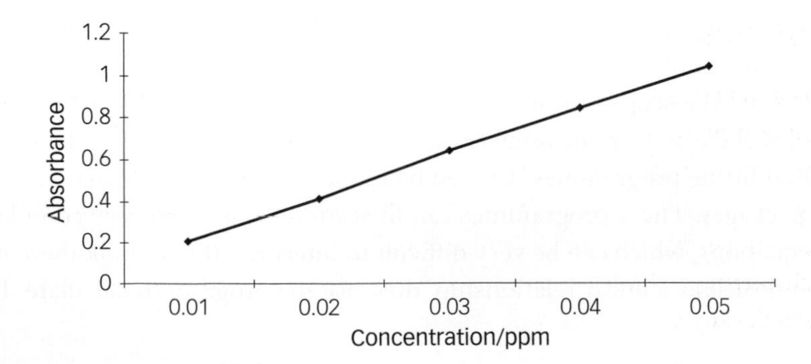

or by calculating the regression line that shows the best fit to the data. You should only plot a best fit line if there is a statistically significant coefficient of correlation. This coefficient value should be given in the figure legend.

Checklist for graph drawing. Make sure that:

- you have used the appropriate form of graph for displaying your data;
- the data are displayed on the correct axes – the independent variable is on the *x*-axis, the dependent variable on the *y*-axis;
- the axis scales are appropriate for the spread of the data;
- the data points and the axis labels are large enough to be clearly legible;
- the axes are labelled and the units of measurement given;
- the graph has a clear figure legend or title so the reader can appreciate what is being displayed without needing to read the accompanying text.

Look at the boxes (p. 125) at the end of this chapter for some examples of how not to draw graphs.

Using images

You might use a variety of images, such as reaction schemes, in your reports. The key rules here are to make sure that:

- the illustration has a legend or title that explains what is being shown;
- images taken using microscopes have a scale bar.

Results checklist

After you have completed the writing of your results section, read it through carefully and ask yourself the following questions:

- Have I clearly described the key findings in the text?
- Do the tables and graphs present the data in a clear form?
- Are the key conclusions supported by appropriate statistical tests or other evidence?
- Have I avoided trying to interpret the findings?

8.2.6 Discussion

The discussion is the section of the report where you interpret the findings from your experiment and place them in the context of the research literature. When it comes to writing up a practical class or project, this is often the most demanding part of the exercise because you must display your understanding of the work you have done and how it fits into the studies undertaken by other scientists.

A useful starting point for the discussion is to write an introductory paragraph that summarizes the key findings from the results:

> In this investigation into the hardness of tap water sample, it has been shown that…

This paragraph then provides a good link with the results section and gives you a clear list of the points you need to interpret. There should also be a link back to the aims that you set out in the introduction.

Having identified your key findings, you then need to discuss each of them in turn. For a research project or open-ended experiment, though probably not for a laboratory experiment, you will need to refer to the relevant research papers, highlighting their findings and relating them to what you have found. It may be that your findings are different from, or even contradict those, presented in other papers. In such cases you will need to explain why you think this may be so: don't always fall back on the staple explanation that you probably got something wrong! Whilst that may be true in many cases, there may also be genuine reasons for your experiments having generated different results. If you are writing up a research project, it is likely that you will have new data and these will need careful interpretation with reference to the research literature.

8.2.7 Conclusions

You should remember to round off the discussion with a concluding paragraph that summarizes both the key findings and their interpretation. In class reports, you may also be encouraged to identify what experiments you might do next in order to progress the work. An example of a short conclusion for our experiment might read as follows:

> In these experiments, it has been demonstrated that there is variation in the hardness of water with geographic region. These findings have been discussed in relation to the studies by Smith & Jones [17] who reported that …..
>
> The accuracy of the titration technique was very good but the precision was poor. This was probably due to the difficulty in reading the endpoints accurately. The atomic absorption method gave good accuracy and precision.

8.2.8 References

As with all your written work, you must include a full list of the references that you cited in the text. See the section on *Citations and references* in Chapter 5, *Working with different information sources.*

8.3 Other sections

You may wish to include some additional sections to the report. The most common of these would be a section on further work, an appendix, and a set of acknowledgements.

Although the results section is where you present your data, this is often in the form of processed information: for example the means of sets of measurements. However, you may wish to include all the raw data, to demonstrate the full set that you have obtained and from which you have derived the summary results. You may also wish to include worked examples of your calculations, or copies of safety assessments, or standard operating procedures. Under such circumstances, these can be presented in *appendices*. This is useful in a project report but would not be typical practice in a research paper.

It is unlikely that you will need to include an *acknowledgements* section in a report from a class practical, but this is more common in a project report. For your project you may have been given specific support by academic or technical staff or have been given access to specific items of equipment that would not normally be available to you. It is then a matter of good practice to acknowledge this help in a brief statement at the end of the report. This should not be treated like the eulogies at the Oscar ceremonies: just give brief, factual acknowledgement of any specific support you were given.

8.4 Write the sections in an appropriate order

Writing a report can often seem a daunting process, particularly if it is for a large piece of project work. Before you put pen to paper (or fingers to keyboard), check again any brief you have been given for the structure of the report so you know exactly what is expected of you.

Some sections of the report should be much easier to write than others. For example, the methods section is simply a factual description of what you did and so this is often a good section to start with.

The next section to tackle is the results. You should think carefully about how you want to present the results first and then produce your schemes, graphs, and tables and carry out any numerical or statistical analyses. When you have the presentation sorted out, the process of writing is fairly straightforward since you are aiming to describe the results that you have in front of you. Don't forget that you are telling a story in which the story-line is illustrated by the figures and tables.

Having set out the key findings of the experiments in the results section, you are then in a position to consider their interpretation. Before beginning writing the discussion, though, you should read through the relevant research papers so that you have a clear picture of the relationship of what you have done to the current research. Take each of your key findings in turn and write about it, making sure you explain the relationship with the previous research in order to interpret what you have found. Your conclusions section pulls together your key findings.

The final sections to write are the introduction and the abstract. Having written the discussion, you should be in a position to present the background to the work and to set out the aims. As for the discussion, you will need to refer to the relevant research articles to support your statements. This should lead logically into the aims of the experiment and you can check off the aims against your final conclusions. The abstract is the final piece to be written, based on a very brief synopsis of the other sections of the report and highlighting the key findings of the research.

Finally, as with all written work, it is always a good idea to put the completed report to one side for a few days and then re-read it, checking that it all ties together and could be understood by someone who had not done the experiments.

* Chapter summary

In this chapter we have highlighted the importance of adhering to accepted structures and style when writing up a practical report or project. Remember, when writing the report it is important to

- use an appropriate writing style;
- adhere to the conventions of scientific report structure;
- read through and check your report to make sure that it tells a logical story;
- check for errors or omissions.

Further reading

Jardine, F.H., (1994), *How to do your Student Project in Chemistry*. London: Chapman and Hall.

Miller, J.N. and Miller, J., (2005), Statistics and Chemometrics for Analytical Chemistry, 5th edn. Harlow: Prentice Hall.

Appendix: examples of how not to draw graphs

The display of data in graphical form is a key part of the preparation of many reports of practical work. In the section on graphs, we have described some of the key principles to drawing graphs but there are many pitfalls, especially when using spreadsheets or graph-drawing packages to draw the graphs. Some of these pitfalls are more obvious than others. Have a look at the graphs shown below as Figures 8.4 and 8.5 and list the errors that you see: each shows at least one error!

Figure 8.4 shows a number of common errors (though rather exaggerated) that might be made when displaying the data similar to that shown in Figure 8.3.

FIGURE 8.4 Plot of relationship between absorbance and concentration.

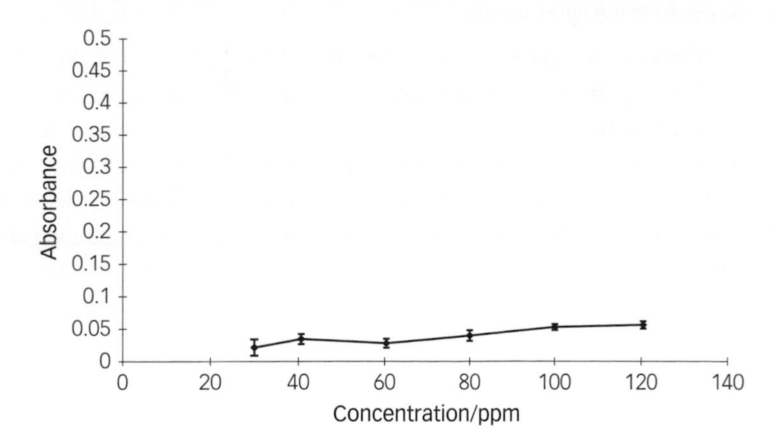

- The figure legend is too brief and does not explain what is being illustrated. The test here is, if the reader only had the figure and the legend to go on and no accompanying text, would he or she be able to understand what the axes indicate.
- The font size for the scale on the *x*-axis is too small to be legible. Make sure that the fonts for the two axes are legible and are the same size.
- The numbers on the *y*-axis are not set to the same level of precision. Make sure that the same number of decimal places is specified for the scale indicators.

FIGURE 8.5 Plot of relationship between absorbance and concentration for the analysis of calcium in water samples.

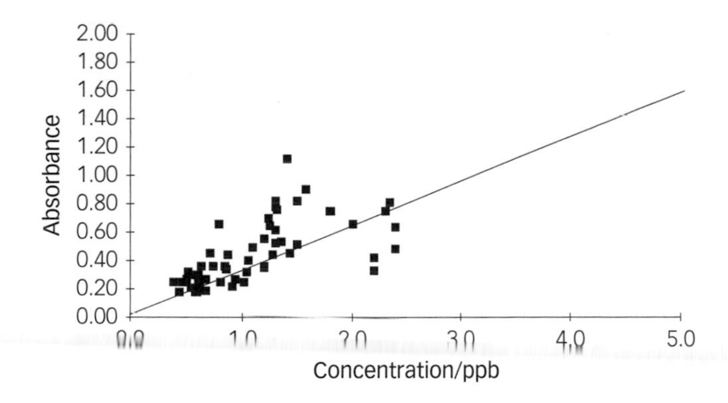

Figure 8.5 at first sight might appear to be okay since the axes are appropriate and clearly labelled and the legend informs the reader as to what data are being presented. However, there are some key flaws in the presentation:

- A straight line has been drawn through the data points but there is no indication as to the basis for drawing such a line: was it simply drawn by eye as appearing to be the line that best fitted the data, or was it calculated using linear regression? The method for arriving at drawing such a line needs to be stated in the figure legend.

- Is it valid to extrapolate or interpolate a line? The straight line here has been drawn back to the origin (coordinate 0,0); this is interpolation and it makes the assumption that the linearity of the relationship continues below the range of the data set on which the calculation was based. You therefore need to ask yourself whether this assumption is justified. In this case, was a blank sample run? The same argument holds true for extrapolation, the process of extending the theoretical relationship beyond the range of the data set. In this case, the Beer–Lambert law deviates from linearity as the concentration increases. The rule of thumb here then is that if you draw a straight line through the data set, it should only extend over the range of the data points and not beyond unless you have strong evidence that the relationship will hold true for other data points.

Chapter 9

Writing for a non-scientific audience

Introduction

As chemists we are generally not very good at communicating the importance of our subject to the general public. Just look at the science pages on the BBC website as evidence for this; lots of stories relating to biosciences, astronomy, physics, and the environment but hardly any about the amazing advances chemistry is enabling our society to make. Think about all of the general science programmes on the television or the general interest science books in the best sellers lists. How many of those are produced by chemists? Not very many. This lack of engagement may contribute to some extent to the poor public image of chemistry and chemists. As an undergraduate you will get plenty of opportunities to write about chemistry for an audience who understand chemistry i.e. your tutors. However, the opportunity to think about communicating effectively with non-specialists is less common. In this chapter we will think about communicating with a range of different people. Although the chapter focuses on written communication the same principles will apply to giving oral presentations too.

9.1 Who is the audience?

If you end up earning your living using your chemistry background it is possible that you may have to communicate the outcomes of your experimental work to important audiences. These people may be senior chemists or research directors who will have a good technical knowledge, probably better than yours. But it is also likely that your audience may contain people who do not have the same technical background as you. These people may include scientists with different specialisms, for example engineers or bioscientists, or they could be people without any technical know-how at all from areas such as sales or finance. Whatever the background of your audience it is important that you can communicate your subject to them in a way that they can understand.

You may have been in the position yourself of having attended a lecture or read a research paper that was far too difficult for you because you did not have the specialist knowledge at that time. You will be aware that it is very difficult to stay focused and interested in such a situation. That is what you need to avoid when presenting material for others.

Below are extracts from two pieces of text. They are both on the loosely related topic of silica. Read the extracts and think about the intended audience and how that has affected the style of writing and presentation.

'Liquid glass, a revolutionary invisible non-toxic spray that protects against everything from bacteria to UV radiation, could soon be used on a vast range of products. The spray, which is harmless to the environment, can be used to protect against disease, guard vineyards against fungal threats and coat the nose cones of high-speed trains, it has been claimed.

The versatile spray, which forms an easy-clean coating one millionth of a millimetre thick – 500 times thinner than a human hair – can be applied to virtually any surface to protect it against water, dirt, bacteria, heat and UV radiation. It is hoped that liquid glass, a compound of almost pure silicon dioxide, could soon replace a variety of cleaning products which are harmful to the environment, leaving our world coated in an invisible, wipe-clean sheen.

The spray forms a water-resistant layer, meaning it can be cleaned using only water. Trials by food-processing companies showed that sterile surfaces covered with a film of liquid glass were equally clean after a rinse with hot water as after their usual treatment with strong bleach.'

'Addition of charged surfactant to mixtures of air, water and hydrophobic silica nanoparticles under high shear induces transitional phase inversion from a water-in-air powder to an air-in-water foam. Optical and electron microscopy reveal the non-spherical shape of both drops and bubbles, respectively, in these materials due to their partial coverage by particles. Complementary experiments are described to elucidate the origin of phase inversion. From surface tension and contact-angle measurements, the ratio of adsorption of surfactant at air/water and solid/water interfaces is determined. Particles become increasingly hydrophilic on adding surfactant since molecules adsorb tail down, exposing charged head groups to the aqueous phase. The increased particle dispersibility into water and the generation of negative zeta potentials confirm this scenario. In addition, the ability of the same silica nanoparticles to act as antifoams of aqueous surfactant foams formed under low shear is investigated. The effectiveness of these particles as antifoams decreases both on increasing the surfactant concentration and the time particles and surfactant are in contact before foaming. Both trends are in agreement with the above-mentioned findings that when particles are rendered more hydrophilic via surfactant adsorption, they no longer break foam films between bubbles but remain dispersed in the aqueous phase.'

The first piece is actually taken from the science pages of the Daily Telegraph. It is still available on their website at http://www.telegraph.co.uk/science/science-news/7125556/Liquid-glass-the-spray-on-scientific-revelation.html (accessed 3rd March 2010). It is obviously aimed at an audience on non-scientists but probably reasonably educated people. The second abstract is taken from a research journal. The article is by Bernard Binks and coworkers (Binks B.P., Johnson A.J. and Rodrigues J.A., Inversion of 'dry water' to aqueous foam on addition of surfactant, *Soft Matter*, 2010, **6**, 126–135) and written not just for other chemists, but other chemists working in the same field. You can see that the language used and the assumptions made in each piece are very different. These are the key factors to consider when writing for a non-specialist audience.

9.2 Vocabulary and symbolism

As chemists we use a language that is all our own. We use names for elements and compounds and names for processes, theories and models. This turns chemistry into a foreign language that is impenetrable for the uninitiated. The selection of appropriate language is essential when writing for a non-expert audience. For example, when writing for another chemist we might write:

'A promising new form of radiotherapy for brain tumours involves the irradiation of boron compounds with low-energy neutrons. Boron neutron capture therapy (BNCT) involves injecting the patient with a ^{10}B-labelled boron compound that preferentially binds to tumour cells. When irradiated with neutrons the ^{10}B undergoes nuclear fission and produces a helium nucleus (an alpha particle) and $^{7}Li^{+}$ nucleus and liberates approximately 2.4 MeV of energy:

$$^{10}_{5}B + ^{1}_{0}n \longrightarrow ^{4}_{2}He + ^{7}_{3}Li^{+}$$

The helium nucleus and lithium ion lose energy by causing ionizing events in the immediate vicinity.' (from *Inorganic Chemistry*, Atkins *et al.*, 2010)

For a non-scientific audience we could remove all the scientific detail and present the basic information much more simply:

A promising new form of radiotherapy for brain tumours has been developed. Boron neutron capture therapy (BNCT) involves injecting the patient with a compound that preferentially binds to tumour cells. When irradiated the compound releases a radioactive substance that destroys cells in the immediate vicinity.

The symbolism that chemists use is even more effective at alienating a potential audience. In the example above the symbols for the different isotopes of the elements and the equation are unintelligible to all but those with a specific education and training. They impart useful information to the specialist, but add nothing for the non-scientists. Compare the following two representations of the same piece of information:

$$CH_3COOH\,(aq) + NaHCO_3\,(s) \rightarrow CH_3COONa\,(aq) + H_2O\,(l) + CO_2(g)$$

and

When vinegar is added to baking powder there is much fizzing as carbon dioxide gas is produced.

The overall message is exactly the same but the first version is inaccessible to most people.

In order to communicate your science to a non-specialist audience you have to understand it much more thoroughly than when you communicate it to an audience of other scientists! For example, in order to explain the principles of your last laboratory experiment to a friend who is studying history you would have to have a good grasp of what you were doing and why you were doing it and be able to convey that into everyday language. A well-known quotation sums it up perfectly:

'If a scientist cannot explain to the woman who is scrubbing the laboratory floor what he is doing, he does not know what he is doing.' Lord Rutherford

And for balance:

'If a scientist cannot explain to the man who is emptying the laboratory bins what she is doing, then she does not know what she is doing.' Tina Overton

Try this: Interpreting equations

This equation explains why ammonium nitrate is an explosive. Explain it in words to a non-scientist.

$$2\,NH_4NO_3\,(s) \rightarrow 2\,N_2\,(g) + O_2\,(g) + 4\,H_2O\,(g)$$

9.3 Engaging your audience

If you browse through the science journalism articles in the newspapers or on their online websites you will see that all the stories have some human interest or a real-life context to draw in the reader. They do not discuss entirely theoretical science, however important and ground breaking it might be. Is this perhaps why chemistry is so poorly covered? Are we not very good at demonstrating the real-life applications of chemical advances? As an author you have to ask yourself why a reader would want to read your article!

We have spent a lot of time in this book talking about the style of writing for the chemical sciences (Chapter 6). The rules that we use when writing for other scientists can be relaxed when we are writing for a different audience. For example, in this extract from an article on kinetic theory from the *Dorling Kindersley Science Encylopedia* you can see that the author has adopted a very informal style in keeping with his intention to make science interesting for young people.

> 'Have you ever wondered why you can smell food cooking? The reason is that tiny gas molecules from hot food whirl through the air and some reach your nose. Although it is hard to believe, the atoms and molecules that make up everything we see are constantly moving. As the temperature rises, the particles move faster, and so they take up more space. This is the kinetic theory of matter.'

In science journalism, the rules of objectivity are often relaxed in the interest of getting a good story. For example, look at these examples of titles from the *Daily Telegraph*'s science pages (http://www.telegraph.co.uk/science/):

> 'Men become hopeless show offs in front of attractive women'
> ' "Smart salad dressing" could save Venice'
> 'Recycling robot sorts rubbish into six piles'

They are attention grabbing and certainly not the sort of titles that a scientist would use for an article. Journalists choose attention-grabbing titles, like those above, and follow the title with an exciting first sentence to draw the reader in. Then they are careful to use only normal language and to avoid the symbolism common to science. Finally, they wrap up with a final message, something for the reader to take away. This approach is quite different from those that we have discussed so far when describing writing for scientists.

Try this: Writing for the press

The local newspaper has announced that is going to run a short series of science based supplements and has decided to begin with a chemistry theme. You have decided, of course, to submit an article. Your article should be based on a recent paper published in the RSC journal *Chemical Communications*. The article must be half a side of A4 and must include one picture or graphic.

✱ Chapter summary

In this chapter we have discussed the importance of being able to communicate chemical sciences to a range of different audiences. The key issues to remember are:

- identify who your audience is;
- identify what your audience already knows;
- take care to us appropriate language and symbolism;
- choose a style that is appropriate.

Chapter 10

Avoiding plagiarism

Introduction

This chapter addresses the increasingly common issue of plagiarism. We will begin by defining what plagiarism is and then look at some of the reasons students give for committing plagiarism in their work. We will then outline a method for you to avoid plagiarism and highlight some of the many good reasons there are for making sure you do. It is important to note that whilst most forms of plagiarism occur in the context of essays and practical reports it is possible to plagiarize in any form of communication; so the study skills required to help you to avoid plagiarism in your writing are essential to your success in every area of your academic work.

10.1 What is plagiarism?

Plagiarism is an increasingly common problem in universities. The increase is partly due to the internet and copying and pasting features of word-processing applications, making copying from one source to another a very easy process. In addition to this, the academic conventions surrounding plagiarism are less widely understood; students often arrive at university with only a very limited understanding of the issue. Because of the rise in the incidences of plagiarism, universities are much more aware of the matter and are therefore keen to address it. Many universities are now using plagiarism-detection software [illegible] recently, which has made it much more difficult for students to get away with plagiarism. However, it is important not to get the issue out of perspective, as we shall see, plagiarism is often not deliberate and there are many positive reasons for learning how to avoid it. But first, we need to make sure we understand what it is we mean by 'plagiarism'.

10.1.1 Definition

The *Oxford English Dictionary* defines plagiarism as:

> The action or practice of taking someone else's work, idea, etc., and passing it off as one's own; literary theft.

So, plagiarism is about using someone else's work but giving the impression that it is your own. Let's think for a moment about the different elements of this definition:

- *taking* – could be from a book, journal, internet site, or even a lecture or handout;
- *someone else's work* – could be someone else's words but could also be their ideas, data, or images;
- *passing it off as one's own* – could be either deliberate or accidental and could be in the context of an essay, a lab report, or a presentation (in fact, any piece of work);
- *literary theft* – tells you in unambiguous terms how it is perceived by many, including your tutors.

So put simply, plagiarism is a form of cheating. You wouldn't cheat in exams and you shouldn't cheat in coursework either. Therefore, just as there are penalties for cheating in exams, there are penalties for cheating in coursework too.

10.1.2 Academic method

Why are universities generally, and academics in particular, so concerned about plagiarism? Why is it considered to be such a big deal? Are they over-reacting or are there good reasons for their concern? Well, let's think for a minute about what underpins academic work. To be more specific, let's consider what is sometimes known as the 'academic method'.

It was Isaac Newton who said of himself (in a letter to Robert Hooke in February 1676):

> If I have seen further it is by standing on the shoulders of giants.

By 'have seen' he was referring to the tremendous discoveries he had made during his lifetime, by 'standing on the shoulders of giants' he was largely referring to the people who had gone before him whose ideas he developed and built upon. In making this statement Newton was highlighting a very important principle of academic study: that whilst you might research and then write-up something individually, you are in fact contributing to a collective endeavour. Even the most novel or original of discoveries will build upon the ideas of others, and it is important that those ideas are acknowledged and that it is only your own contribution that you claim for yourself.

So, plagiarism is an important issue because academic honesty underpins the academic method: it is fundamental to all academic work. If you think about it, this also means that if you can learn the study skills that enable you not to plagiarize, you will be helping yourself not only to avoid being penalized, but more positively, to perform well in all areas of your academic work.

10.1.3 **Deliberate plagiarism**

When people hear the word 'plagiarism', they usually think of someone deliberately copying text without referencing it, but as we have seen from the definition above, plagiarism is often much broader than simply copying text. We will deal first with examples of deliberate plagiarism.

Copying

Imagine a student used the following text in an essay on 'Alternative energy sources'.

> There are possible stores of fossil fuels on the ocean floors. Methane clathrates are crystalline solids formed at low temperatures when ice crystallizes around CH_4 molecules. Clathrates are also referred to as *methane hydrates* or *natural gas hydrates* and their formation has caused major problems in the past by clogging gas pipelines in cold climates.

The first sentence is the student's own but the next two sentences are not written by the student; rather they are taken from *Inorganic Chemistry* by Atkins *et al.* (2010). The student has therefore copied the text without any indication that some of the words have come from another source. They have therefore plagiarized by copying.

FIGURE 10.1 Sample of a plagiarized image.

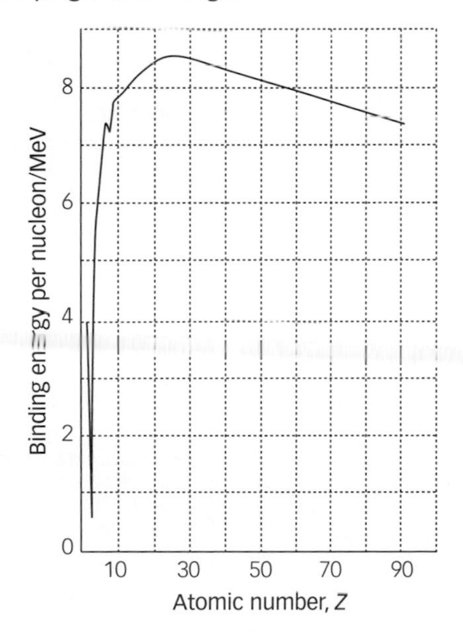

Alternatively, take an example of an image: imagine a student had used Figure 10.1 in an essay on 'Nuclear chemistry'. If the student failed to acknowledge the source of the image, in this case *Inorganic Chemistry* (2010) again, they would also be plagiarizing by copying. It is important to remember that plagiarizing by copying images is equivalent to, and so just as serious as, plagiarizing by copying of text.

Collaborating

It is also possible to plagiarize by collaborating. Collaboration is a form of copying: you copy from a fellow student (who you might be working with during a practical class, for example) rather than copying from a published source. This is often seen by students as a legitimate practice, but it is important to remember that whilst you will often be encouraged to work with a fellow student to conduct a practical experiment, this doesn't usually mean that you are being asked to collaborate on the writing-up aspect of the report as well. Obviously there will be similarities between two reports written by two students who had worked together to conduct an experiment, but these will largely be confined to the methods and results sections. However, in the discussion section, the students will be expected to analyse and draw conclusions from their findings *independently* of each other. For example, compare the two discussion sections in Figure 10.2 from students A and B who were lab partners for a practical class on analysing the chloride content of groundwater.

You can see that there are definite similarities between the two discussions; the marker would probably conclude that the students had collaborated on the write up of the discussion section and had therefore plagiarized.

FIGURE 10.2 Collaboration in lab write-ups.

Student A

Discussion

In these experiments, the chloride analysis of the ground water was carried using both titrimetric and gravimetric analysis. As expected the results were similar in each case. The concentration of chloride ions in the sample was found to be 156 ppm by titration and 189 ppm by gravimetric analysis. The gravimetric value may be higher because of incomplete drying of the sample. The end point for the titration was quite distinct so there could be more confidence in that value. The gravimetric analysis was very time consuming to carry out required more skill that the titration.

Alternative method

An alternative method for chloride analysis in ion selective electrode 3. The advantages of the ISE are

Student B

Discussion

Chloride content of groundwater was determined using titration and a gravimetric analysis. As expected, both methods gave similar results: the titration gave 156 ppm, gravimetric analysis gave 189 ppm. The gravimetric method gave a higher value, probably due to insufficient drying of the sample. The titration may be the better method because it gave a very distinct end point and was quick and easy to carry out. The gravimetric analysis took a long time and a lot of skill to carry out.

I would recommend that ion selective electrode could be an alternative method. This would be a good method because it is quick and easy, accurate and can be carried out *in situ*.

FIGURE 10.3 Independent lab write-ups.

Student A

Discussion

In these experiments, the chloride analysis of the ground water was carried using both titrimetric and gravimetric analysis. As expected the results were similar in each case. The concentration of chloride ions in the sample was found to be 156 ppm by titration and 189 ppm by gravimetric analysis. The gravimetric value may be higher because of incomplete drying of the sample. The end point for the titration was quite distinct so there could be more confidence in that value. The gravimetric analysis was very time consuming to carry out required more skill that the titration.

Alternative method

An alternative method for chloride analysis in ion selective electrode 3. The advantages of the ISE are

Student C

Discussion

A sample of groundwater was analysed for chloride ions. The two methods that were selected were a) titrimetric analysis with silver nitrate solution and b) gravimetric analysis by precipitation of silver chloride by addition of silver nitrate solution. Both methods gave similar results; 156 ppm and 159 ppm respectively. These results are very similar indicating that the methods are reliable. A comparison of the two methods suggest that the titration method is easier to perform and is rapid, whilst the gravimetric method is very time-consuming. I would recommend that of the two methods the titration is most appropriate.

A method of analysis which would be better for industrial applications is the ion selective electrode. This would be better because

Now look at the examples of students A and C in Figure 10.3.

Whilst we would still expect to see similarities in the methods and results sections, we can see that the students have thought independently about their findings and come to their own conclusions. This doesn't mean (in this case) that they have come to different conclusions, but the wording and sentence structure suggests that they have thought for themselves.

Self-plagiarism

It is also possible to self-plagiarize, this is when a student submits their own work, which they have previously submitted for another part of their course, but fails to declare it had been used for such a purpose. This is probably the type of deliberate plagiarism that students find the most frustrating; 'I'm the author of the work' you might reason, 'so surely I can submit it as my own work for whatever aspect of my course I choose.' Whilst it is difficult not to have some sympathies with this view, self-plagiarism is considered unethical because of the deception involved in submitting the same material for credit in different modules or courses. On a more positive note, writing can always be improved upon, so it is in your own interest to do more than just recycle previously submitted work; it also helps the learning process. This is in addition to the more obvious point that a good answer written for one essay question, for example, is probably not going to be specific enough to be considered a good answer to a different, even if related, essay question. In short, whilst disallowing self-plagiarism may seem a little harsh, it is to your advantage to tackle each piece of assessed work as a separate exercise, allowing you to tailor your answer to the specific task and broaden your knowledge of the subject.

Buying

It is possible, on an apparently increasing number of websites, to buy essays, even complete dissertations, for submission as assessed work. This is the most blatant and outrageous form of plagiarism there is. Just in case you are ever tempted to do this, here are a number of good reasons why you shouldn't:

- it's expensive;
- it's fraudulent;
- you learn nothing about the subject;
- the quality is seldom good;
- it will be easy for your tutors to spot if there is a significant disparity between a bought essay and your usual standard of work.

Don't do it!

10.1.4 Accidental plagiarism

Some plagiarism is deliberate but more commonly it is accidental, or at least less deliberate than the examples cited above. There are lots of reasons for accidental plagiarism; we deal with some of them below, and use them to form the basis for the following section, *How to avoid plagiarism.*

Ignorance

Not that you will have this excuse after reading this chapter, but ignorance is the most basic reason. 'I didn't know' is a common response to accusations of plagiarism, particularly from undergraduates in the early stages of their degree course.

Poor planning

Poor planning or failing to manage your time well may seem unrelated to plagiarism, but they are, in fact, frequent causes of it. Most students would be shocked at the idea of deliberate plagiarism, especially in a pre-meditated manner, but when you are running out of time to complete an essay the temptation to take short-cuts becomes stronger. Read the case study below and see if you can appreciate what we mean.

Case study: Steve

Steve was busy with another assignment and then went out to celebrate when it was finished. Waking up in the morning he remembered that he had a second assignment to hand in and so quickly used Google to search for some information on the web. Using a mixture of copying, pasting, and rephrasing he put together his assignment in a couple of hours. However, his tutors spotted what he had done and he was given a mark of zero.

So, plagiarism as a result of poor planning is more plausible than it may at first seem.

Failure to record source details

Failure to record source details is probably the simplest form of plagiarism. It starts at the note-making stage; you find an appropriate piece of text, image, or data that you would like to reproduce or adapt in some way for your essay or report, but you don't write down where the material came from. At the essay drafting stage you either forget you got it from somewhere else and believe it to be your own original work, or you realize it came from elsewhere but either can't be bothered or don't have the time to re-trace the source, so you use it anyway without referencing and hope that no one will notice. Does this seem implausible? Read the following case study and see if you understand what we mean.

Case study: Andria

Andria took a lot of notes while she was researching her assignment, but didn't pay much attention to noting down the referencing details of her sources. When she wrote her assignment she ended up very unsure about what were her own ideas and what she had taken from books. Her lecturer is now talking about plagiarism but she feels that she was only a bit disorganized and is horrified that she is being accused of dishonesty.

Plagiarism due to failure to record source details is, therefore, a very real problem.

Inappropriate notes

Inappropriate notes are also a common cause of plagiarism. It is reasonably straightforward to record source details (in theory at least) but how do you know what sort of notes to make on the source? See if the following case study helps you understand what we mean.

Case study: Abdul

Abdul has a good understanding of the course; however, he finds it difficult to put things into his own words. Academic textbooks always seem to put things so well that he hasn't got much to add. Sometimes it just seems easier to copy out chunks of different books, paraphrase them a bit, and organize them into an answer. He doesn't feel that this should be considered plagiarism as he has researched the material and put the assignment together from a range of different sources. Unfortunately his tutors don't agree.

Academic text books or journals can indeed 'put things so well' that it's difficult for a student to know what to add. This is why it is so important to work at understanding the concept that you are dealing with, and it takes a certain amount of confidence to do this. We will deal with this in the section *How to avoid plagiarism*.

Incomplete citing and/or referencing

Sometimes students inadvertently commit plagiarism by simply not recording their citations and references fully. This could be due to a lack of understanding of what is required, or a lack of willingness to add what are sometimes just considered to be 'finishing touches', but are in fact crucial to the integrity of their work. Again, see if the following case study helps illustrate this.

Case study: Jasmine

Jasmine worked hard on her essay and was pleased with what she produced. However, she has never understood the point of the referencing system. She included all of the books that she used in a reference list at the end of the essay but didn't bother with citing her sources in the body of the essay. She thought the reference list at the end would be enough but, unfortunately, her lecturers viewed this lack of citing as plagiarism. They covered the assignment in red ink and gave her a very low mark.

So, whilst sorting out the citations and references might just seem like an optional extra, it is vital to get it right. Referencing correctly will be dealt with in Section 10.2.5 on p. 149, and more detailed guidance is available in the section on *Citations and references* in Chapter 5, *Working with different information sources*.

Lack of engagement with sources and failure to express own ideas

The final reason students might plagiarize that we are going to deal with, and possibly the one that requires the most thinking to address, is lack of engagement with source material and failure to express your own ideas. This is where a student may well have planned ahead, recorded source details, made appropriate notes, and cited and referenced accurately, but still doesn't demonstrate that they have thought a great deal about what they have written. It is perfectly possible, in an assessed piece of work to simply string together lots of other authors' ideas (either quoted or paraphrased), but if you fail to include your own ideas, in the form of comment or analysis on the material, you still might be plagiarizing, even if the source material is cited and referenced correctly.

10.2 How to avoid plagiarism

We have dealt with what plagiarism is, having thought about its definition and how academic honesty (or not plagiarizing) un derpins the academic method. We have also considered the main reasons students plagiarize; identifying both deliberate and accidental forms of plagiarism. We are now going to consider how to avoid it.

10.2.1 Know what plagiarism is

The first step to avoiding plagiarism is to be able to recognize it when you see it. This is different from simply defining it; can you tell in practice when something is or is not plagiarized? Use the following exercise to test your knowledge so far (exercise adapted from Willmott, C. and Harrison, T. (2003).[1]

Essay extract exercise

Original text:

'Methane clathrates are crystalline solids formed at low temperatures when ice crystallizes around CH_4 molecules. Clathrates are also referred to as *methane hydrates* or *natural gas hydrates* and their formation has caused major problems in the past by clogging gas pipelines in cold climates.'

Essay extract 1:

Methane clathrates are crystalline solids formed at low temperatures when ice crystallizes around CH_4 molecules. Clathrates are also referred to as *methane hydrates* or *natural gas hydrates* and their formation has caused major problems in the past by clogging gas pipelines in cold climates.

Is this extract plagiarized, yes or no?

Essay extract 2:

Methane clathrates are crystalline solids formed at low temperatures when ice crystallizes around CH_4 molecules. Clathrates are also referred to as *methane hydrates* or *natural gas hydrates* and their formation has caused major problems in the past by clogging gas pipelines in cold climates. [1]

Is this extract plagiarized, yes or no?

Essay extract 3:

'Methane clathrates are crystalline solids formed at low temperatures when ice crystallizes around CH_4 molecules. Clathrates are also referred to as *methane*

Willmott C.J.R. and Harrison T.M. 2003. An exercise to teach bioscience students about plagiarism, *Journal of Biological Education*, **37**, 139–140 (Thanks to the *Journal of Biological Education* for permission to adapt this article.)

hydrates or *natural gas hydrates* and their formation has caused major problems in the past by clogging gas pipelines in cold climates.' [1]

Is this extract plagiarized, yes or no?

Essay extract 4:

Clathrates are formed when ice crystallizes around small molecules. The most common clathrates are methane clathrates and which can form in natural gas supplies and 'their formation has caused major problems in the past by clogging gas pipelines in cold climates'. [1] Other small molecules that form clathrates include ethane and propene.

Is this extract plagiarized, yes or no?

Essay extract 5:

Methane clathrates are formed at low temperatures when ice crystallizes around CH_4 molecules. They are also known to as *methane hydrates* or *natural gas hydrates* and they have caused major problems in the past by clogging gas pipelines in cold climates. [1]

Is this extract plagiarized, yes or no?

Essay extract 6:

Clathrates are crystalline solids that are formed when water crystallizes around small molecules such as methane, ethane and propene. When clathrates are formed in natural gas pipelines they can cause problems as they can clog the pipelines. [1]

Is this extract plagiarized, yes or no?

Essay extract feedback

Essay extract 1:

This is copied word-for-word but there is no reference. It is a definite case of plagiarism.

Essay extract 2:

This is marginally better than extract 1 because the source has been acknowledged by the inclusion of the reference. However, it is still plagiarism: the text is copied word-for-word and should be in quotation marks to indicate that it is the exact wording of the source.

Essay extract 3:

Quotation marks are used to acknowledge that words have come from a different source. The quotation marks here make it clear that the student is acknowledging that both the ideas and the words have come from the textbook, so it is not plagiarized. However, stringing together a series of quotations does not demonstrate your understanding of the subject and so is likely to score low marks.

Essay extract 4:

The quotation is indicated and used in an appropriate way; it has been commented on by the author and supplemented with further information, and so is not plagiarized.

Essay extract 5:

This is only a cosmetic alteration. The wording and sentence construction bears very close resemblance to the source and so is plagiarized.

Essay extract 6:

The student has understood the source and put the relevant information into his or her own words. This demonstrates the student's engagement with the text and ability to explain the information relevant to the essay, and so is not plagiarized.

Try this: Identifying plagiarism

Decide whether any of the students below are guilty of plagiarism or collusion.

Katy has trouble with a part of a mechanism and asks Ahmed for help. Ahmed shows his own answer to Katy to demonstrate how it has been done.

Eric has trouble with a part of the experimental write-up and asks Jason for help. Jason types in some of the answers for Eric.

James looks at Li's solution to a problem without Li knowing, in order to see some workings that may help him with his tutorial.

Julie has trouble with a project and finds some material on the Internet. She copies it, makes a few changes, and hands it in.

10.2.2 Plan ahead

We have already noted that bad planning, and the pressures it creates, increases the temptation to plagiarize. So whilst bad planning is not a direct cause of plagiarism, it is still nonetheless, a significant factor, and therefore needs addressing.

The solution to bad planning is, of course, good planning. This strikingly obvious statement is worth mentioning, as it highlights the fact that the skills required to address this issue are simply common sense. However, there can be a world of difference between knowing what you should do and actually doing it. In Chapter 7, *Writing essays and assignments*, the different stages involved in writing an essay were dealt with. It is helpful to think of these stages not just in terms of separate tasks but also as distinct periods of time that need planning. Chapter 7 recommended that you think about the following stages when writing an essay:

- analyse the question and brainstorm ideas;
- research the topic;
- write a plan;
- write first draft;
- review and re-draft;
- proofread (including checking of citations and references).

A good way to plan ahead is to order these tasks into some sort of timeline, calendar, or diary. Figure 10.4 shows an example timeline representing a period of three weeks that you might have available to write an essay. It is worth spending a short amount of time allocating your tasks in this way to provide some structure and discipline to your time and so hopefully reduce the chances of you leaving things to the last minute, which would, as we have seen, increase the chances of you plagiarizing.

This method also has the advantage of making the task less daunting. If you had on your 'to do' list; 'write fullerenes essay' it would probably be a long time before you were able to cross it off, which can be quite demotivating. However, if your 'to do' list said things like 'gather journal information for essay' and 'write essay plan', these are things that would take hours rather than weeks so you would be able to cross things off much sooner. This makes it easier to discern progress and so is motivating rather than demotivating; you feel like you are getting somewhere.

Note also in Figure 10.4 that there is a significant gap between 'write first draft' and 'review and re-write first draft'. This is important for a couple of reasons: firstly, you could reward yourself in some way for getting this far in the process, again a motivating factor; and secondly, coming back to an essay after a bit of a break enables you to be more objective about what you have written. This objectivity increases the likelihood that you will be able to make significant improvements.

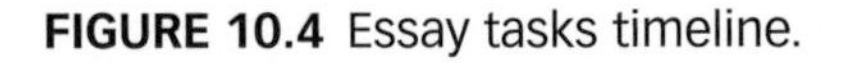

FIGURE 10.4 Essay tasks timeline.

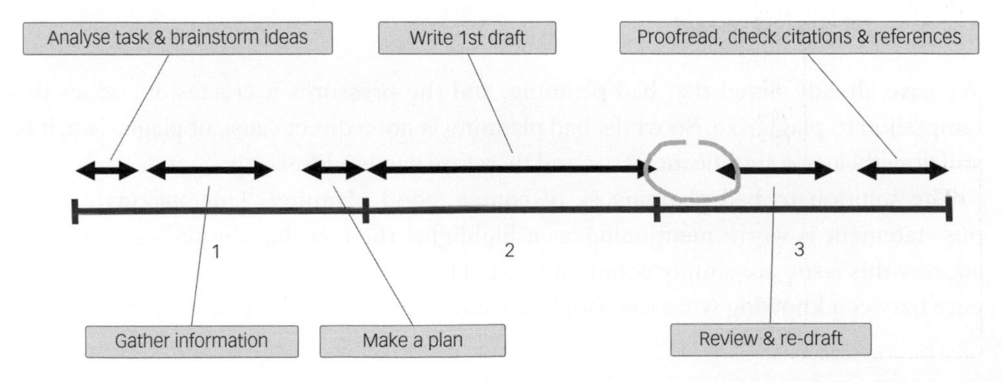

10.2.3 **Record source details**

As we have seen, it is important to record source details if you don't want to plagiarize. It sounds simple enough, but as with planning ahead, when the pressure is on it is both easier and more tempting to make mistakes.

Imagine you had made this note during a study session in the library last week:

Fullerene production is a simple process. A large electric current is passed through graphite rods in a quenching atmosphere of inert gas, thus evaporating the rods to produce a light, fluffy condensate called 'fullerene soot'.

It is a good piece of writing and perfect for your essay on fullerenes. The only trouble is you can't quite remember if these are your words or someone else's because you didn't write down any more information; an easy mistake to make.

The words are actually taken from a chapter by Lamb in *The Chemistry of Fullerenes* (2003). So if you were to use them in your essay without putting them in inverted commas (' '), citing and referencing them, you would be plagiarizing. The good news is that, because failure to record source details is a simple mistake to make, there is also a simple solution. It just takes a bit of discipline.

There are four main sources that you will use in your studies: books, journals, websites, and lectures. Table 10.1 summarizes the information you need to record for each one in order to reference fully.

TABLE 10.1 Source details to record for different information types.

Source type	Source details required
Books	Name(s) of author(s)* Year of publication Title of book Name of publisher Place of publication Page reference *If it's a chapter of an edited book you also need to include the 'Editor(s) name(s)', 'Title of chapter', and 'First and last page numbers of chapter'.
Journals	Name(s) of author(s) Year of publication Title of paper Title of journal Volume number First and last page number
Websites	Name(s) of author(s)* http:// address Title Date page accessed* *Often internet pages indicate when they were last updated—this is also important information to record if available.

TABLE 10.1 Cont'd

Source type	Source details required
Lectures	Name of lecturer Name of lecture/lecture series Date of lecture Location (e.g. University of Leicester)

All you need to do is make sure you record the source details relevant to the type of source before you begin to make notes. It's as simple as that. See Table 10.2 for examples of citations and references from the source details.

TABLE 10.2 Examples of source details recorded.

Source type	Source details required	Example source details	Example reference
Books	Name(s) of author(s): Year of publication: Title of book: Name of publisher: Place of publication: Page reference:	Paul R. Jenkins 1992 Organometallic Reagents in Synthesis OUP Oxford pp.95–96	Jenkins P R, *Organometallic Reagents in Synthesis*, OUP, Oxford, 1992
Journals	Name(s) of author(s): Year of publication: Title of paper: Title of journal: Volume number: First and last page number:	Kroto H.W., Heath J.R., O'Brien S.C., Curl R.F. and Smalley R.E. 1985 C_{60}: Buckminsterfullerene Nature 318 162–163	Kroto H.W., Heath J.R., O'Brien S.C., Curl R.F. and Smalley R.E., C_{60}: Buckminsterfullerene, *Nature,* 1985, **318** 162–163
Websites	Name(s) of author(s): http:// address: Title: Date page accessed:	World Energy Council http://www.worldenergy.org/focus/fuel_cells/default.asp Fuel cells 1/03/10	World Energy Council, Fuel cells [online] available from: http://www.worldenergy.org/focus/fuel_cells/default.asp [accessed 1st March 2010]
Lectures	Name of lecturer: Name of lecture/series: Date of lecture: Location:	Tina Overton Introduction to organometallic chemistry 23/11/10 University of Hull	Overton T., Introduction to organometallic chemistry 2010 [lecture 23rd November 2010]

10.2.4 Make appropriate notes

Recording source details, then, is quite straightforward, as long as you know what details you need to record for which source. But what about when you actually need to make notes on the source material? How do you do it?

When taking notes it is important to ask yourself a simple question: what am I taking notes for? There are several possible reasons you might have. You might be taking notes to:

- select information;
- understand information;
- remember information.

Clearly these three areas overlap considerably, but in order to keep things simple we are going to deal with them separately. Chapter 5, *Working with different information sources*, deals with making notes to understand information and the reading techniques necessary to achieve this efficiently; Chapter 14, *Getting the most out of revision*, addresses taking notes to remember information. Here, we will just focus on taking notes to *select* information.

At the beginning of this chapter, when we were thinking about the different elements involved in the definition of plagiarism, we identified that 'someone else's work' could be their words but also their ideas, data or images. When making notes, therefore, we are not merely thinking about text, we also need to remember other forms of information too: including ideas, data, and images. In simple terms, you have two options when considering how to include another person's work in your own: you could reproduce it exactly (so in the case of text this would mean quoting it word-for-word) or you could adapt it in some way (again, in the case of text this would mean putting it into your own words, i.e. paraphrasing it).

For example, if you were writing an essay on 'The principles of catalysis' and found some useful information in the textbook *Inorganic Chemistry* by Atkins *et al.* (2010), you may decide that the exact wording and/or imagery is important to you and so want to reproduce it exactly, as follows in the text and Figure 10.5.

'Two stringent tests of any proposed mechanism are the determination of rate laws and the elucidation of stereochemistry. If intermediates are postulated, their detection by magnetic resonance and IR spectroscopy also provides support. If specific atom-transfer steps are proposed, then isotopic tracer studies may serve as a test. The influences of different ligands and different substrates are also sometimes informative.' [1]

FIGURE 10.5 The determination of catalytic mechanisms [1].

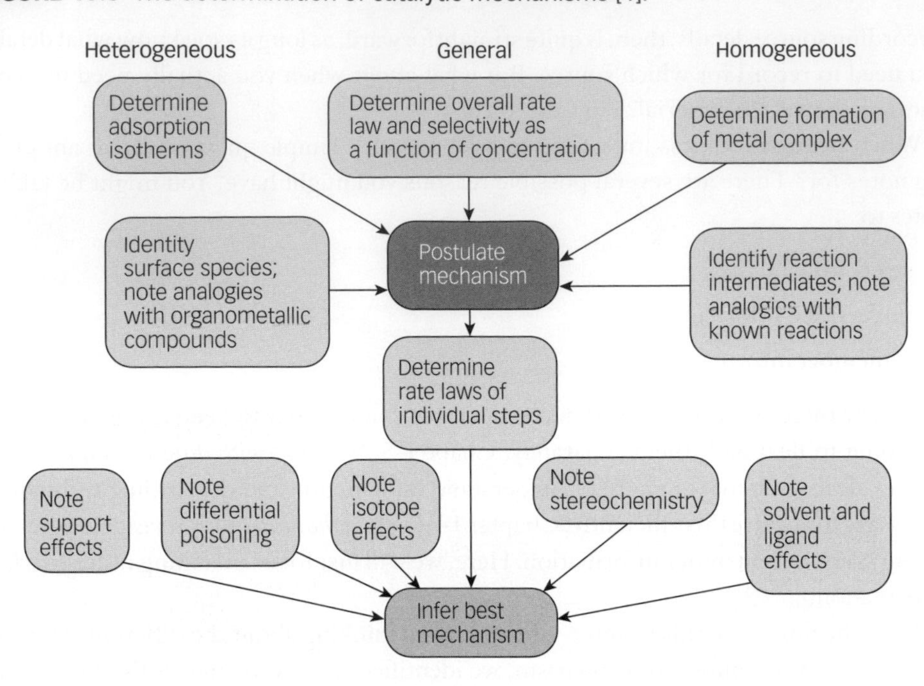

However, if it was the idea rather than the exact wording or imagery that was important, you would need to adapt it in some way (which may mean simplifying or expanding on it), as follows as in the text and Figure 10.6.

When investigating a catalytic mechanism, the rate laws, stereochemistry, and the nature of ant intermediates are important. NMR, IR, and tracer studies are useful in elucidating the structure of any intermediates [1].

The important thing to remember is that, whether you are reproducing another's work (e.g. quoting text) or adapting it (e.g. paraphrasing text), both need referencing. In the case of quoting, the quoted text is placed in inverted commas (' ') (to make clear that you are using someone else's words verbatim) followed by the citation, in the case of paraphrasing the paraphrased text is followed by the citation.

10.2.5 **Reference correctly**

As with planning ahead and recording source details, referencing correctly is simple enough in theory, you just need to pay attention to the details. Here are some simple steps that should help (more detailed guidance is available in the section on *Citations and references* in Chapter 5, *Working with different information sources*).

FIGURE 10.6 The determination of catalytic mechanisms (adapted from [1]).

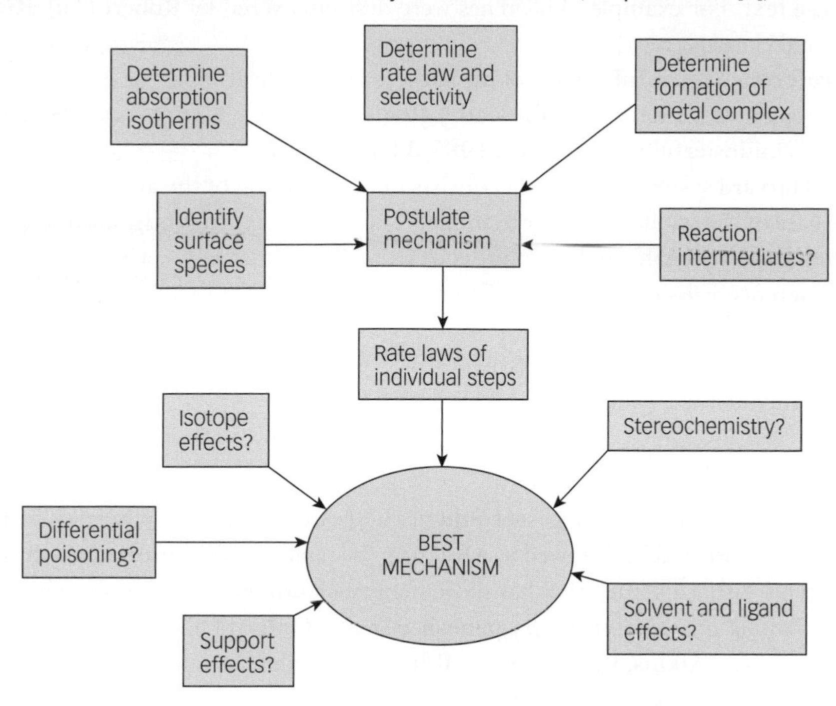

Check what style of referencing is expected from you

Different institutions have different styles of referencing; the most common system in chemical sciences is the Vancouver system adopted by the Royal Society of Chemistry and the American Chemical Society. The Harvard author-date system (see Chapter 5) may be favoured by some tutors and some will use footnotes, some will use endnotes, some will use variants of these systems, and there are many others besides! Even within an individual institution (e.g. a university) it is unlikely that all departments use the same type of referencing style. In fact, there are likely to be minor variations even within departments. This can create a lot of confusion and it can be very tempting to decide not to bother with the details. However, when it comes to referencing, the details are important.

Know the difference between citing and referencing

An important distinction to make is the difference between citing and referencing. In general terms **referencing** refers to the practice of acknowledging your sources. More precisely, however, referencing in an academic context has two elements:

1. the citation in the text;
2. the full reference at the end of the piece of work (e.g. your essay or practical reports)

The **citation** is simply the way you refer to another's work in your own work, in the case of the Vancouver system that is, the text, idea, image or data is followed by number of the reference. The citations start at number one and are then numbered consecutively within the text. For example, 'Fullerenes were first discovered by Robert Curl, Richard Smalley and Harold Kroto [1]'.

The **reference** is the full details of the authors and publication at the end. See Figure 10.7. For example, '1. Kroto H. W., Heath J. R., O'Brien S. C., Curl R. F. and Smalley R. E., C_{60}: Buckminsterfullerene, *Nature,* 1985, **318,** 162–163'.

In the Harvard system the citation consists of the surname of the author (or authors) and the date of the publication. For example, 'The fullerenes are synthesized by passing an electrical discharge through graphite rods (Kroto *et al.*, 1985)'.

The reference is listed at the end. The references are listed in alphabetical order by author name.

Kroto H. W., Heath J. R., O'Brien S. C., Curl R. F. and Smalley R. E., C_{60}: Buckminsterfullerene, *Nature,* 1985, **318,** 162–163.

Use *et al*. correctly

'*Et alii*' is a Latin term meaning 'and others'. It should therefore be written in italics (because it's Latin) and be followed by a full stop (because it's an abbreviation). It is used when you are citing a source that has more than two authors, for example, when referring to the book *Inorganic chemistry* published by the Oxford University Press, instead of having to write Atkins, P., Overton, T., Rourke, J., Weller, M., and Armstrong, F., you would simply write Atkins *et al.*, which, as you can see, is very handy.

FIGURE 10.7 The relationship between citations and references.

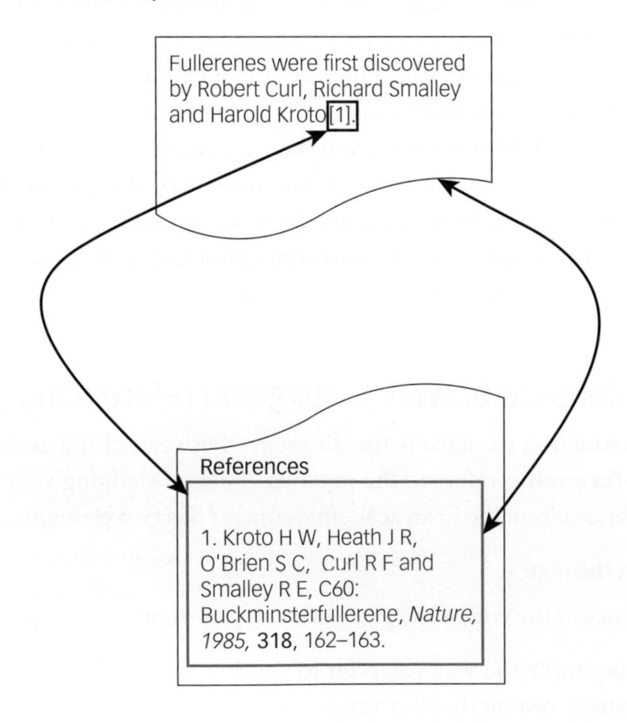

List the references correctly

If using the Vancouver system the references are listed in numerical order at the end. For example:

1. Dalgleish, R., 2002 ...
2. Wells, C., 2004 ...
3. Harrison, T., 1999 ...

If using the Harvard system list the references at the end of your document be in alphabetical order by surname of the author, for example:

Dalgleish, R. 2002 ...
Harrison, T. 1999 ...
Wells, C. 2004 ...

If there is more than one author for a publication then it is the surname of the first author that counts, for example:

Cann, A. 2005 ...
Jones, M. and Adams. C. 2003 ...

The Harvard system requires extra care. If there are two authors with the same name or one author with two publications, then you need to also order the list by date, for example:

Dalgleish, R. 2002 ...
Dalgleish, R. 2004 ...

If an author has more than one publication in a given year then the list needs to distinguish between these, usually by including letters in the dates, for example:

Robbins, H. 2004a ...
Robbins, H. 2004b ...

This distinction would also need to be reflected in the citation, so (Robbins, 2004a), and so on.

Consider using bibliographic software, if available

If you are going to be using a large number of references, for instance, in a final-year dissertation, you may wish to use bibliographic software. Bibliographic software is designed to allow you to input, organize, manage, retrieve, and format lists of references more easily than manual input. Two examples are EndNote and RefWorks, but it will depend on what software your university subscribes to. Whatever system your university subscribes to it is worth finding out more about it as it may save you a lot of time. However, even if you use such software, you still need to understand the structure of references if you are going to use it appropriately.

10.3 Seven good reasons not to plagiarize

As we have seen, avoiding plagiarism is not just about correct references, it is an unavoidable part of good academic writing. In case you need any more convincing, here are seven good reasons for you not to plagiarize.

1. Plagiarism short-circuits learning

When you are set an assessed piece of work it is easy to focus on the tangible endpoint, that is, the mark you are awarded. However, your tutors set you work not merely as a means of allocating marks, but also as a means of helping you to learn. The research you conduct, the selecting of information, and the process of writing all aid the learning process. The more you plagiarize, the less you learn.

2. Plagiarism destroys independent thinking

Your tutors are interested in what you think about a subject. Obviously, your point of view needs to be supported by appropriate evidence, but as we have seen, that evidence needs to be accompanied by your own comment or analysis; your own ideas and conclusions. If you plagiarize by not commenting or analysing evidence, you are not thinking for yourself; you are merely communicating the ideas of others and have failed to think independently.

3. Plagiarism is unfair to your peers

How would you feel if a student on your course scored a higher mark than you on an essay but you knew that they had cheated? Make sure you don't make anyone else feel the same way about you.

4. Plagiarism is a form of theft

Words, ideas, data and images that have been written, thought of, discovered, or created by someone else belong to that person. Using them without acknowledging the source is stealing.

5. Plagiarism can result in severe penalties

Different universities have different penalties for students who have plagiarized. The type of penalty depends on the precise nature of the plagiarism, the extent to which the work contributes to final marks and whether or not it is a first offence. Common penalties include:

- an awarding of zero for the piece of work, the requirement to repeat the work plus an official warning;
- an awarding of zero for the module and withdrawal of the right to re-sit;
- the downgrading of degree class by one division;
- expulsion from the course.

6. Plagiarism is now more difficult to get away with

Many universities are now using plagiarism-detection software. Students are required to submit their work electronically, the software then compares the work with a vast array of other sources, including core texts, journal articles, internet sites, and previously submitted student work. A report is then produced highlighting the areas of similarity with the compared sources (Figure 10.8). The tutor would then interpret the report to decide whether or not it is plagiarized. The software is becoming increasingly sophisticated and makes getting away with plagiarism much more difficult. However, even if your university doesn't currently operate an electronic detection system, it is worth noting that academics are also very capable of quickly spotting plagiarism using more old-fashioned methods.

FIGURE 10.8 Example of plagiarism software originality report generated by TurnItIn.

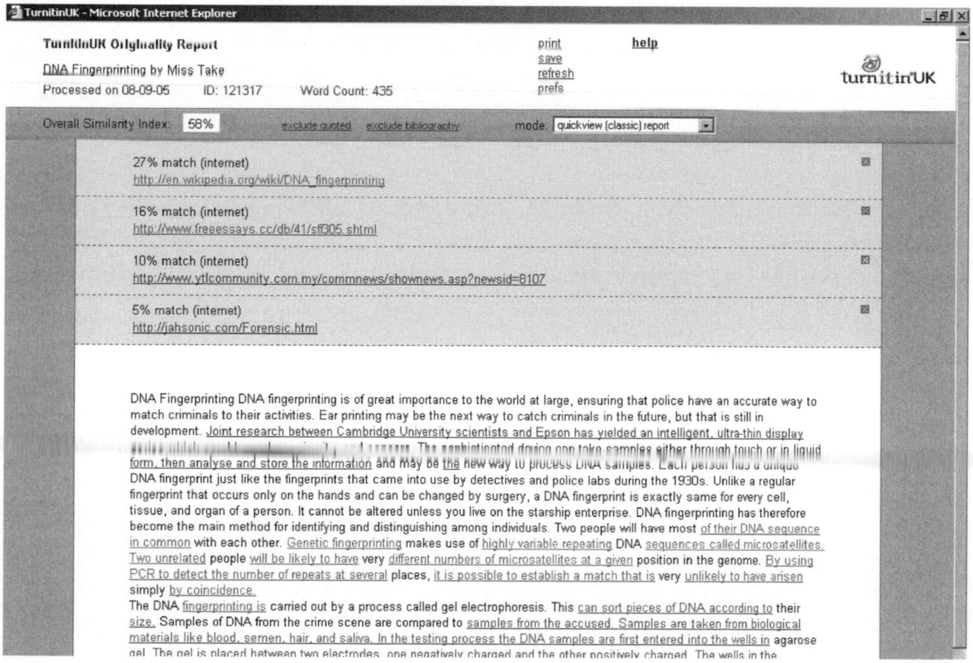

TurnitinUK - Microsoft Internet Explorer

TurnitinUK Originality Report

DNA Fingerprinting by Miss Take

Processed on 08-09-05   ID: 121317   Word Count: 435

print
save
refresh
prefs
help

turnitinUK

Overall Similarity Index: 58%   exclude quoted   exclude bibliography   mode: quickview (classic) report

27% match (internet)
http://en.wikipedia.org/wiki/DNA_fingerprinting

16% match (internet)
http://www.freeessays.cc/db/41/sff305.shtml

10% match (internet)
http://www.ytlcommunity.com.my/commnews/shownews.asp?newsid=8107

5% match (internet)
http://jahsonic.com/Forensic.html

DNA Fingerprinting DNA fingerprinting is of great importance to the world at large, ensuring that police have an accurate way to match criminals to their activities. Ear printing may be the next way to catch criminals in the future, but that is still in development. Joint research between Cambridge University scientists and Epson has yielded an intelligent, ultra-thin display [illegible] through touch or in liquid form, then analyse and store the information and may be the new way to process DNA samples. Each person has a unique DNA fingerprint just like the fingerprints that came into use by detectives and police labs during the 1930s. Unlike a regular fingerprint that occurs only on the hands and can be changed by surgery, a DNA fingerprint is exactly same for every cell, tissue, and organ of a person. It cannot be altered unless you live on the starship enterprise. DNA fingerprinting has therefore become the main method for identifying and distinguishing among individuals. Two people will have most of their DNA sequence in common with each other. Genetic fingerprinting makes use of highly variable repeating DNA sequences called microsatellites. Two unrelated people will be likely to have very different numbers of microsatellites at a given position in the genome. By using PCR to detect the number of repeats at several places, it is possible to establish a match that is very unlikely to have arisen simply by coincidence.

The DNA fingerprinting is carried out by a process called gel electrophoresis. This can sort pieces of DNA according to their size. Samples of DNA from the crime scene are compared to samples from the accused. Samples are taken from biological materials like blood, semen, hair, and saliva. In the testing process the DNA samples are first entered into the wells in agarose gel. The gel is placed between two electrodes, one negatively charged and the other positively charged. The wells in the

7. Not plagiarizing means your learning will be more effective

Finally, remember that avoiding plagiarizing is not simply about evading punishment. The skills that help you not to plagiarize are the skills that enable you to be an effective learner. Learning these skills will help you in all areas of your academic work; helping you to understand your subject better and so help improve your grades.

* Chapter summary

As we have seen, plagiarism is an important issue, and you need to know both what it is and how to avoid it if you are going to do well in your studies. In order to avoid plagiarizing you should:

- plan your work so that you do not run out of time;
- record details of all source materials;
- make appropriate notes on sources;
- re-write information from sources in your own words;
- cite sources appropriately;
- reference correctly and consistently;
- do not be tempted to cut and paste from websites;
- collaborate with peers but do not copy or collude.

Sources

Atkins, P., Overton, T., Rourke, J., Weller, M., Armstrong, F., 2010. *Inorganic Chemistry*. Oxford University Press.

Lamb, L.D., 2003. in *The Chemistry of Fullerenes* ed. Taylor R., Singapore: World Scientific Press.

Chapter 11

Preparing scientific presentations

Introduction

The ability to present information verbally to a group of people is an increasingly important skill for students undertaking chemical science degrees, or any other kind of degree for that matter. Some students find the prospect of standing up in front of a group of their fellow students and academic staff to give a presentation is the stuff of nightmares, even more so when the presentation is assessed. For other students presentations are not quite such a frightening prospect, but even so, the public nature of a presentation, as compared to the privacy and anonymity of an essay or exam for example, is a factor that makes getting presentations right all the more important; no one likes being embarrassed in front of their peers.

In this chapter we will begin by considering the purpose of presentations as a form of assessment; we will then identify some of the characteristics of effective presentations, and scientific presentations in particular. We will look at the importance of finding out the answers to a number of key questions before progressing onto analysing the brief, researching the topic and planning, and ordering your presentation's content. Finally, we will highlight the importance of preparing effective and appropriate notes. This chapter complements Chapter 12 that looks specifically at delivering, as opposed to preparing for, your scientific presentation.

11.1 What are presentations supposed to achieve?

It can be useful to think about why you are being asked to undertake a particular form of assessment, as different forms of assessment are chosen for different reasons. The assessment of presentations is usually divided into two broad areas: content, and the communication of the content (or 'substance' and 'style'). Both of these elements are important and you need to be careful not to focus on one at the expense of the other. For example, it is possible for your presentation to have relevant and informative content, but if you don't communicate it in a well-structured and engaging manner you would score relatively low marks in an assessment. Equally, you should avoid trying to dress up content that you don't understand and haven't put much effort into researching by giving a slick presentation with a few impressive PowerPoint slides. If you do this it is usually fairly obvious to the audience (and it certainly will be to your tutor) that your presentation is all style and no substance. So, both substance and style are important.

We will consider the following brief for a sample presentation:

Prepare and deliver a five-minute oral presentation on the high-temperature superconductors. The presentations will take place at 12 noon on Friday 1st March in room 101.

We can identify a list of the skills being exercised in order to complete the task as follows:

- using feedback from previous presentations to produce a better piece of work;
- analysing the question or title (and associated instructions);
- researching the topic;
- bringing together the information gathered from a range of sources, e.g. lecture notes, textbooks, research papers, web articles, etc.;
- planning the information to address the brief and make sure that all the key points are covered;
- matching the level of detail against the type of audience you will be addressing;
- ordering the material in a logical manner;
- creating visual aids to support your key points;
- preparing appropriate notes so that you can remember what you want to say;
- practising your presentation beforehand so that you become familiar with communicating the content;
- delivering your presentation in a relaxed and engaging manner;
- referencing the information correctly to acknowledge the original sources;
- planning and managing your time to make sure everything is ready by the deadline.

As was similarly identified in Chapter 7, *Writing essays and assignments*, this is a surprisingly long list and could probably be extended. The important point is that while some of these skills, such as researching the topic and preparing your notes, are obvious and easily recognized as part of the process, others, such as using feedback and referencing information, are more implicit but still very important if you are to plan and deliver a good presentation. As we go through the process of preparing a presentation (this chapter) and delivering a presentation (Chapter 12), we will consider these different skills and give you guidelines for improving the quality of your scientific presentations.

11.2 What makes a good presentation?

What, then, makes a good presentation? Nearly everyone has an opinion on this question. Even if you have never given a presentation before, you will have seen others give presentations (for instance a lecture), and you will probably remember some that were really good, and some that were really bad! It is useful to think about what makes a good presentation as it will give you a list of things you want to try to copy and things you want to try to avoid.

11.2.1 Characteristics of effective presentations

Obviously, not all presentations should be the same. In fact, it is important that as you gain experience in presenting you develop your own style, rather than trying to conform too rigidly to a set standard. However, effective presentations usually exhibit certain characteristics. These include that the presentation is:

- relevant;
- informative;
- well structured;
- engaging.

Whilst clearly this is not an exhaustive list, these are important characteristics of effective presentations. You should also recognize that these characteristics are not self-contained; they overlap and are dependent on each other. To go back to our substance and style distinction that we mentioned above, relevance and informativeness are largely to do with substance, whilst structure and engagement are largely to do with style. We will deal with each of these characteristics in turn.

Relevant

The starting point for any presentation is the presentation brief, as it is this that will govern the content and focus of the presentation. Whilst it can be tempting to jump straight from a brief to the bits you find interesting about the topic or the bits that you happen to have found some useful information about, it is important that you keep going back to the title and instructions you have been given to make sure that your material is relevant, just like you should do for an essay. But a presentation should not only be relevant with regard to the brief; it should also be relevant with regard to the audience. Are you pitching your presentation at the correct level for the audience in terms of their current understanding and their ability to assimilate the material? For instance, is it likely that you will be using quite technical terms on occasions? If so, are these terms your audience is already familiar with, or do you need to define them so that they know what you are talking about? So, a presentation needs to be relevant, and relevant to both the brief and the audience.

Informative

Not only should a presentation be relevant, it should also be informative. What is the content that you want to communicate to your audience? What do you actually want to say? It can be very helpful during your preparation to try to summarize the content of your presentation in a single sentence or three or four short bullet points. For instance, to go back to our example above, you could summarize the content of that presentation as follows: 'Short explanation of what high-temperature superconductivity is, what compounds are superconducting, applications of superconducting materials.' If you can't summarize the content in a sentence or two then it probably isn't focused enough. So, a presentation needs to be informative.

Well structured

Now we start to get on to the style, as opposed to substance, elements. The structure of a presentation is a good example of where there is significant overlap between style and substance: good structure begins with the content. However, you need to be able to communicate that structure to the audience so they can follow what you are saying. If a presentation is easy to follow, it will be due in part to it having a good structure. There is some much-quoted but sound advice for structuring a presentation that is often attributed to the advice of Aristotle. It boils down to this:

> Tell 'em what you're gonna tell 'em. Tell 'em. Tell 'em what you've told 'em.

There are two simple ways of communicating the structure of your presentation to your audience: you could tell them or you could show them (or both).

Telling the audience the structure involves outlining it verbally at the beginning, for instance:

This presentation is divided into three main sections. First, I'll give a brief explanation of what high-temperature superconductivity is. Then describe the structures of some of the compounds that exhibit high-temperature superconductivity. Finally, I'll tell you about some real applications.

In addition to outlining the structure of the presentation at the beginning, it will also be helpful to give verbal cues to the audience as you go along so they can clearly follow the progression of the presentation. Verbal cues act like signposts to the audience so they know where you are going. Some of these verbal cues include:

- a single word such as 'First', 'Secondly', and so on;

- linking statements such as 'So that's the background to the technology, let's now think about its application…';
- concluding statements such as 'In conclusion…' and 'Finally…';
- finish with something like 'Thank you for listening'.

That might all seem fairly obvious to you, but it is surprising both how often these signposts are neglected and how helpful it is for the listener when presenters make a point of using them. As you become more experienced in presenting you will give these verbal cues much more naturally, but while they don't come naturally it is a good idea to make a conscious point of using them.

Showing the audience the structure could simply be a case of producing an outline slide that summarizes the main points. An example is shown in Figure 11.1. You could use this at the beginning when you tell the audience about the structure. If the structure was particularly important or particularly complex, you might also repeat the outline slide within the presentation when you mention the verbal cues. However, it is possible to give a presentation so much structure that it becomes very separated out and doesn't flow, so be careful not to overdo it.

Engaging

The last characteristic of an effective presentation that we have identified is that it should be engaging; that is it is interesting and holds the audience's attention. Again, there is overlap here with the content or substance aspects of the presentation (particularly the relevance to the audience), but whether or not a presentation is engaging is much more to do with how information is communicated as opposed to what is communicated. Whether or not a presentation is engaging will depend on a wide range of factors; it will need to be well paced, the presenter will need to have good eye contact with the audience, and it will probably help if the presenter can be fairly relaxed (or at least seem to be relaxed!). All of these depend a large amount on how much you have practised the presentation. An engaging presentation is often also helped by the appropriate use of visual aids, which we will address in Chapter 12.

FIGURE 11.1 Outline slide showing presentation structure.

Outline

- What is high temp superconductivity?
- What type of compounds exhibit it?
- Applications
- Examples
  - Wires, trains, Hadron collider
- Conclusion

11.2.2 Characteristics of effective scientific presentations

In addition to the characteristics of effective presentations highlighted above, there are certain characteristics that scientific presentations in particular should exhibit. These include that a scientific presentation should be:

- objective;
- precise;
- clear.

Again, this is not an exhaustive list and there is overlap between the different elements, but they are key characteristics nonetheless. We will address each of them, briefly, in turn.

Objective

Science is, by its very nature, objective. That is to say, observations made during scientific research should be impartial and not influenced by personal feelings or opinions. So it follows then that a scientific presentation should also be objective. This means that the content you are communicating needs to be well researched and well referenced. As was highlighted in Chapter 10, *Avoiding plagiarism*, whilst students are usually aware that plagiarism is wrong in the context of essays, there is often a bit of a blind spot when it comes to presentations. But making sure that you reference the sources of information in your presentation is not simply about avoiding plagiarism; just as importantly it is about showing the objectivity of what you are saying.

Precise

Scientific presentations should also be precise. As was highlighted in Chapter 7, *Writing essays and assignments*, the use of generalizations often clouds the meaning of what is written or indicates a lack of confidence on the part of the presenter. For example, what does the following phrase actually mean?

Superconductivity occurs in certain materials at fairly low temperatures…

The phrases 'occurs in certain materials' and 'fairly low temperatures' have no real meaning as their interpretation depends entirely on the listener. Relative terms such as 'quite' or 'fairly', or phrases such as 'In general terms…' lack precision and should, wherever possible, be replaced with more precise phrases:

Superconductivity occurs in many metals, alloys and doped semiconductors at very low temperatures, typically below 77 K…

Clear

The final characteristic of scientific presentations that we have identified is that they should be clear. This does not just apply to simple, uncomplicated subjects. In fact, one of the marks of a good presentation and a good presenter is the ability to explain complex things in relatively simple terms. Using simple terms should not mean that the presentation is simplistic, but it does mean that the content needs to be explained in such a way that an average member of the audience can understand it without too much difficulty.

11.3 Analyse the question or brief

We have already identified that relevance is a key characteristic of effective presentations. We observed that a presentation needs to be both relevant to the question or brief you have been set and relevant to the audience you will present to. A useful technique for analysing the question or brief is simply to write it in the middle of a piece of paper and then annotate it, as can be seen in Figure 11.2.

This technique enables you to begin to get your thoughts down on paper quickly whilst encouraging you to link your thoughts to the task you have been set, thereby helping you to keep your ideas relevant. It also allows you to make connections between ideas and see, at a glance, how the presentation might be structured.

FIGURE 11.2 Analysing the question/title.

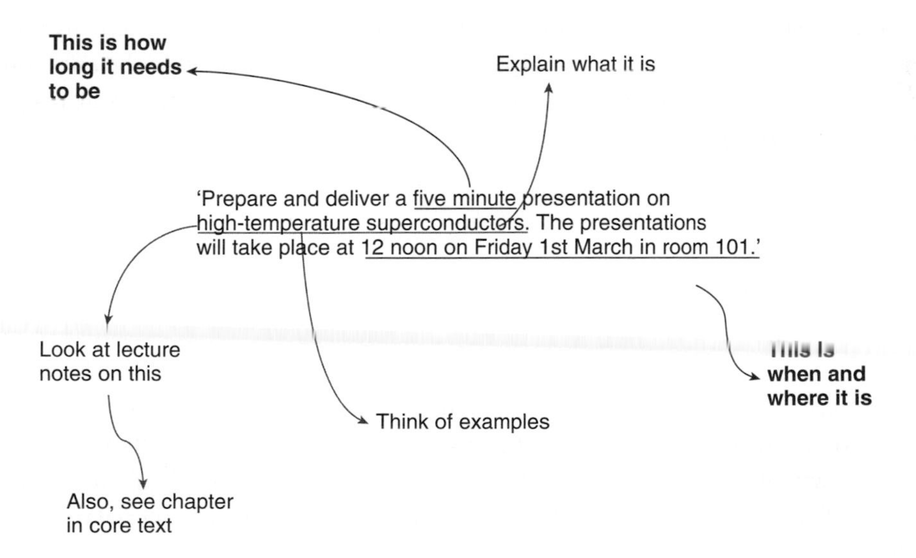

11.4 Research the topic

As we saw in Chapter 7, *Writing essays and assignments*, researching the topic can be a very time-consuming task. There are three things you can do to reduce, or at least control, the amount of time the research part takes.

- Analyse the brief first so that you have some clear ideas of what you are looking for before you start looking for it. It is easy to waste a lot of time researching material that you later discover is not particularly relevant; analysing the brief provides a focus for your research.
- Set a time limit for your research. If you don't set a time limit, the research element can just go on and on, especially if you are being particularly diligent or are very interested in the topic.
- Look at source material in an appropriate order. If you start off with an internet search it can take you off at all sorts of tangents. If, however, you start with lecture notes and key chapters from core texts, it provides focus for your research so it is easier to discriminate when you get to the other sources (see Chapter 5, *Working with different information sources*). A suggested order is as follows:
    - start with the lecture notes – check that you understand the information;
    - identify the essential reading from the relevant chapters in the core texts;
    - when you have mastered the basics extend your reading by looking at books that address a particular aspect of your topic;
    - look at recent research articles to bring you up to date on the subject and to add the necessary level of detail.

It is important to note that jumping to internet sources too early can actually hinder your research. You need to first establish a clear focus that is relevant to the question or brief so that you can be highly selective if you do choose to supplement your research with internet sources.

11.5 Plan and order the material

During the research stage you will have probably gathered more material than you actually need, even if you have been quite disciplined in your approach. The temptation is to include too much material because you have spent the time finding it; however, this will often result in a presentation that is difficult to follow, unfocused, or too long (or all three!). The following stages will help you to decide what material you should include.

11.5.1 Choose your main points

The starting point for selecting and organizing your material should be to choose your main points. You might not be able to list all of the main points straight away, but it is important nonetheless to have a good attempt. As you continue in your preparation you may well come back to your main points to refine or reorganize them, but you need a basic framework at an early stage to focus your thoughts and provide a structure for the rest of your material. An example structure for a five-minute oral presentation on high-temperature superconductors was given above, as follows:

- what is high-temperature superconductivity?
- what materials exhibit high-temperature superconductivity;
- examples: levitation, wires, trains, Hadron collider;
- conclusion.

This is a very sketchy outline, but it is enough information to get you started. A useful technique to help you draft your main points is to simply put your research notes to one side, then try to write your main points without looking at the research notes, expressing each point in a few words or a short sentence. Alternatively, you could try to articulate your main points to a friend, because expressing ideas out loud can be a good way of helping you decide whether or not they are making sense. It may also be helpful to add some notes to your outline to give you an idea of what the focus of your presentation should be. For instance, from the list above it looks like the four bullet points (or at least the first three) are equally sized sections, which probably isn't your intention. It might be better, therefore, to annotate your main points with a bit more information, as follows:

- what is high-temperature superconductivity? (define the properties, mention temperature, i.e. not really very high!);
- what sort of compounds exhibit high-temperature superconductivity, examples, structures;
- applications: making use of levitation, trains, also wires, Hadron topical;
- conclusion (summarize the main points and invite questions).

This is also the point at which you would begin to organize your main points into a logical sequence, so that each point links to, and builds on, the previous one. The sequence of your main points will be determined not only by what is logical though, but also by what you are aiming to achieve. For example, if you are trying to build an argument you will want to move from background information to precise points of detail, or if you are explaining a process you might begin by explaining its purpose and then take your audience through each stage of the process step by step. So our oral presentation on high-temperature superconductivity might go something like this:

What is high-temperature superconductivity? Superconducting materials have a conductivity of zero and have no interior magnetic field. This is called the

Meissner effect. It was discovered by Heike Kamerlingh Onnes in 1911. So it has been known about for a long time but only recently reaching the stage that it is useful.

What you need to keep in mind is that you are trying to communicate information in manageable chunks, helping the audience to understand the progression of your argument or process. It doesn't matter at this early stage if you can't sequence your main points precisely (for instance, is it better to define the effect first or introduce the amazing properties) because such a level of precision is not necessary at this stage. The important thing is to simply choose what your main points will be and make a good attempt at the sequencing; you can always revise it later on.

11.5.2 Select supporting information

We have already identified that there is a temptation to cram too much information into a presentation, simply because you have found lots of information at the research stage. An important arbiter that we have referred to twice previously in this chapter is the relevance of the material (to both the brief and the audience). However, there are other criteria that are helpful to consider explicitly when you get to the selecting supporting information stage. The other criteria that we will consider are clarity, authority, and interest.

Clarity

When deciding whether information should be left in or taken out, ask yourself this question; 'Is this information that I'm thinking of including going to make what I am trying to say any clearer?' Of course, this relies on you knowing what it is you are trying to say, which is why establishing the main points first is important. For instance, in your research you may have found out lots of information about the quantum-mechanical origins of superconductivity, but is letting your audience know this going to help you make what you are trying to say any clearer? If you have decided that the focus of your presentation is the applications of superconductivity then you will probably decide that the underpinning physics is unnecessary.

Authority

The second criterion to consider when selecting supporting information is authority. Ask yourself this question: 'Is this information that I'm thinking of including going to make what I am trying to say any more authoritative?' The authoritativeness of information you use will depend largely on the source of that information; whether or not it is well researched, impartial and up to date. This is usually easier to establish for publications such as textbooks and articles published in research journals than for web resources, especially freely editable sources such as Wikipedia. The authoritativeness of sources is addressed in Chapter 5, *Working with different information sources*.

Interest

Finally: interest. Ask yourself this question; 'Is this information that I'm thinking of including going to make what I am trying to say any more interesting?' Perhaps you have found a picture, chart, or even a video clip, which will help illustrate a point. Anything that will bring what you are saying to life.

So, establishing whether information you have found will make your presentation clearer, more authoritative, or more interesting is a helpful means of deciding whether or not it should be included. But always come back to the bigger, more fundamental question, 'is this relevant?' Also remember that, when you decide to leave information out, it doesn't necessarily mean that the information is wasted: the information you leave out may have been helpful in leading you to other information. It may be useful supplementary information if you are asked questions about your presentation, or it may be useful if you subsequently have to write an associated essay or report.

> **Try this: Adding interest to a presentation**
>
> Find three things that you could use in the presentation of superconductors to add interest. These could be interesting examples of applications, pictures, video, etc.

11.6 Prepare your notes

Once you have analysed the brief, researched the topic, and planned and ordered your material, you are ready to prepare your notes. You have two main options available to you here: you could write them out word-for-word (known as *verbatim*) or you could use outline notes (e.g. bullet points). Alternatively, you could use a combination of these two types. We will consider each of these two options in turn and then look at how they can be combined to best effect.

11.6.1 Verbatim notes

Verbatim notes have many advantages and disadvantages.

Advantages of verbatim notes

Verbatim notes have the obvious advantage that you start the presentation knowing exactly what you are going to say. Knowing exactly what you are going to say can be reassuring and so boost your confidence. Verbatim notes can be particularly helpful, therefore, if you are lacking confidence in a particular presentation:

- perhaps because your first language isn't English;
- or you are presenting information that you are relatively unfamiliar with;

- or you are presenting complex material that requires a level of precision that would be difficult to achieve from outline notes;
- or you have some other reason for feeling nervous.

Disadvantages of verbatim notes

Verbatim notes do have significant disadvantages though. There is a danger that you will simply read out your presentation and therefore won't communicate it effectively. When you look up from your notes you might lose your place in your notes. Also, a read-out presentation is difficult to bring 'alive' because:

- you will lose eye contact with your audience;
- your voice will probably become more monotone;
- you will have much less energy and enthusiasm about the delivery;
- as a result your presentation will be less engaging.

How to use verbatim notes

Verbatim notes seem to have got themselves a bad name. Often people will tell you not to use them because of the very real disadvantages listed above, but this advice usually comes from people who have become experienced presenters and so have forgotten about much of the anxiety that is often associated with presenting. However, as can be seen from the advantages, there are very real benefits too. The trick is to use verbatim notes in the right way. We will take each of the potential disadvantages in turn and suggest a strategy to help you avoid that particular pit-fall.

There is a potential danger that you will simply read out your presentation

Don't! Reading out the presentation will nearly always result in the problems listed. You don't need to memorize your notes but you do need to have practised them sufficiently so that you can work from them as a **prompt**, as opposed to as a **script**.

If you look up from your notes you might lose your place

Quite possibly. So make sure you format your notes in such a way so as to help you not lose your place. For instance, increase the font size, increase the line spacing, use paragraph breaks, use bullet points, use annotations, and highlight key words. Figure 11.3 shows verbatim notes that would be very easy for you to lose you place with; Figure 11.4 shows helpfully formatted verbatim notes that dramatically reduce the chances of you losing your place.

A read-out presentation is difficult to bring 'alive' because you will lose eye contact with your audience, your voice will probably become more monotonous, you will have much less energy and enthusiasm about the delivery, and as a result your presentation will be less engaging.

Again, most of the above will be helped by practice. If you can get to the stage where you have practised your presentation sufficiently so that your notes are a prompt rather than a script, you should be able to think about the communication, as well as content

FIGURE 11.3 Unhelpfully formatted verbatim notes.

High-temperature superconductors

Hello my name is Sam and today I'd like to talk to you about high-temperature superconductivity.

My presentation is divided into four main sections. These are what is high-temperature superconductivity? what sort of compounds exhibit high-temperature superconductivity, I'll give you some examples and structures. Then I'll tell you about some applications that make use of high-temperature superconductivity. Then I'll pull it all together for you with some conclusion.

Well, what exactly is high-temperature superconductivity. Well, ..superconducting materials have a conductivity of zero and have no interior magnetic field. This is called the Meissner effect and can be seen illustrated in this diagram here. Superconductivity was discovered by the Dutch physicists Heike Kamerlingh Onnes in 1911. So it has been known about for a long time but only recently reaching the stage that it is useful. high-temperature superconductors are superconducting above about 30 degrees Kelvin, so they are not really very high temperatures at all. 30 degrees Kelvin was thought to be the highest possible temperature for superconductivity predicted by BCS theory, which stands for Bardeen, Cooper, Schrieffer. Modern interest is in high-temperature superconductors and in trying to increase the temperature below which materials exhibit superconductivity. The Nobel prize for physics in 1987 was awarded to two physicists from IBM who produced the first high-temperature superconductor. The aim these days is to produce materials which are superconducting above the boiling point of liquid nitrogen which is 77 K or -196 °C, so still pretty cold!!

So what sort of compounds exhibit this amazing property? Well, ..

issues – issues such as making eye contact with the audience, varying the tone of your voice appropriately, and generally trying to be more engaging.

11.6.2 Outline notes

Outline notes also have advantages and disadvantages.

Advantages of outline notes

Having your notes in only outline form can free you up to focus on engaging with the audience, rather than focusing on your notes. Also, if you are not reading your notes out, your voice and manner will probably be more natural, which will also help engage your audience. There is no temptation to read your notes out verbatim, which, as we have highlighted above, causes problems.

FIGURE 11.4 Helpfully formatted verbatim notes.

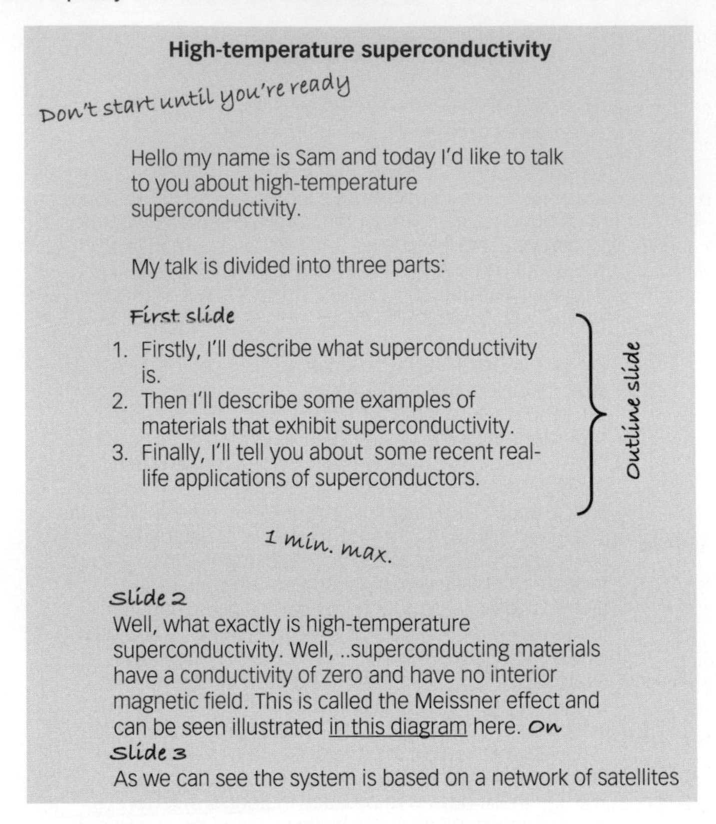

Disadvantages of outline notes

There are two main disadvantages of outline notes. Firstly, you might be more nervous about your presentation because you have fewer notes to rely on. Secondly, you may find that your mind 'goes blank' during your presentation because your outline notes aren't a sufficient prompt to remind you of what you intended to say.

How to use outline notes

As we have seen with verbatim notes, outline notes have both advantages and disadvantages; the trick is to use outline notes in the right way. The only way around these potential disadvantages is practice: you need to rehearse your presentation a sufficient number of times so that you know your outline notes are full enough to remind you of what you need to say. Of course, while you are testing the adequacy of your notes in this way you will also be learning the material. Remember though, just because you can get your presentation right once or twice in the privacy of your bedroom doesn't mean that everything will be fine in the more pressured situation of a tutorial class or assessed presentation.

11.6.3 **Combine note types to maximize benefits**

Given the potential disadvantages of each notes technique, probably the best method to employ is a mixture of the two. Where you are confident that you only need short bullet points to prompt you, you can use outline notes, for instance, for the introduction, as shown in Figure 11.5.

FIGURE 11.5 Outline notes for introduction.

Introduction

- Me
- Title of presentation
- What I'll cover (slide 1)

However, where the material is more complex and precision is more important, you can use at least fuller, if not verbatim, notes, as shown in Figure 11.4.

11.6.4 **Openers and closers**

In addition to planning the introduction, main points and conclusion of your presentation, it is also important to think about how you will open the presentation and how you will close it. It is a good idea, regardless of what type of notes you choose to use, to know exactly what your opening statement is going to be (the 'opener') and exactly what your closing statement is going to be (the 'closer'). If you have a faltering start, perhaps tripping over your opening words or not saying them clearly, it is likely that you will lose confidence and composure, which can be very difficult to then regain. Opening statements need only be simple sentences such as:

- 'Hi, my name's Sam and my presentation is about high-temperature superconductors'.
- 'Hi, my name is Alex and I've got five minutes to talk to you about…'.

Simple statements such as these get you off to a positive start and make it clear to the audience what you are doing. Equally, closing statements are important too. Too often presentations finish with a mumbled apology: 'Err, that's it, I'm finished…' then the presenter sits down. A much better way to finish is a clear conclusion followed by a closing statement such as:

- 'So, in this presentation we have seen the exciting potential of high-temperature superconductors but have also seen that there are still some specific limitations that need to be addressed. Thank you very much for your attention and I will be happy to try to answer any questions.'

11.6.5 A note about working from visual aids

An alternative to speaking from either verbatim or outline notes is to work from a visual aid. This is when you use a visual aid (for example a PowerPoint slide or an object) to prompt what you are going to say, rather than your notes themselves. This can work well because it frees you from your notes entirely yet can still give you sufficient prompts to remind you of what you want to say. However, as with verbatim and outline notes, it all depends on how the method is used. The method will work badly if your slides are essentially copies of your verbatim notes (as shown in Figure 11.6) and you end up reading them out: your audience can always read faster than you can speak and so will quickly get bored. However, if you either condense the bullet points to key words or phrases (as shown in Figure 11.7), or use an image to illustrate what you want to say, you will find it much easier to talk from (this will be covered in more detail in Chapter 12).

FIGURE 11.6 A difficult-to-work-from slide.

What is high-temperature superconductivity?

- Superconducting materials have a conductivity of zero and have no interior magnetic field.
- This is called the Meissner effect.
- Superconductivity was discovered by the Dutch physicists Heike Kamerlingh Onnes in 1911.
- High-temperature superconductors are superconducting above about 30 degrees Kelvin.
- This was thought to be the highest possible temperature for superconductivity predicted by BCS theory.

FIGURE 11.7 An easier-to-work-from slide.

What is high-temperature superconductivity?

- Conductivity of zero
- No interior magnetic field.
- The Meissner effect.
- Discovered by Heike Kamerlingh Onnes in 1911.
- High-temperature superconductors have $T_C > 30$ K.
- Thought to be the highest temperature predicted by BCS theory.

11.7 Practice

Finally, in this section on preparing for presentations we need to highlight the importance of practice. How effective a presentation is will depend a large amount on how much you have practised it. Practising a presentation is important because if you know your presentation well it frees you up to think about the communication, as well as content issues (style as well as substance).

We noted above in *How to use outline notes* that just because you can get your presentation right once or twice in the privacy of your bedroom it doesn't mean that everything will be fine in front of an audience. However, the value of practising in your own environment is not to be under-estimated, and there are a number of ways you can make it more effective.

11.7.1 Practise against the clock

It can be very difficult to judge how much time your presentation will take without actually practising it out loud and against the clock. Depending on the style of notes you have used, sometimes you will look at your notes and think that it will barely last two minutes, never mind five. Other times you will think you have got far too much material but in reality you will get through it far quicker than you thought. The only way to have a reasonably accurate idea of how long your presentation will take is to practise it out loud. Of course, in practising it out loud you are not only getting an idea of how long it takes, you are also learning the material and refining its delivery.

11.7.2 Practise in front of a mirror

In a similar way to it being difficult to know how long the presentation will take it is also difficult to know how you will look. The only way to get an idea of this is to practise the presentation in front of a mirror or, better still, a video camera. You will probably feel a bit self-conscious doing this but it will give you some very useful feedback about how you are presenting yourself as well as how you are presenting your material. It is a particularly helpful way of assessing whether or not you are using your body language appropriately (see Section 12.3.3).

11.7.3 Practise in front of your friends

Most usefully, you could practise in front of some willing friends. This could give you feedback on your presentation's length, your body language, the structure, in fact all aspects of your presentation. It can be a bit embarrassing and you will doubtless make mistakes, but as long as your friends are going to give some constructive comments, and you can deal with the potential embarrassment, it will give you invaluable feedback and so improve your final presentation, and thereby its grade, significantly.

* Chapter summary

The ability to present information verbally to groups of people is an increasingly important skill. It is of particular relevance to scientists as it is an important means of communicating research. This chapter began by identifying the characteristics of effective presentations and then identified the stages necessary to prepare presentations. You should:

- analyse the question or brief;
- research the topic;
- plan and order the material;
- select supporting information;
- prepare your notes;
- practise your presentation.

This brings us on to the next chapter, *Delivering scientific presentations*.

Chapter 12

Delivering scientific presentations

➲ Introduction

You may be wondering why we have devoted an entire chapter just to the delivery of scientific presentations. Chapter 11 dealt in some detail with the preparation aspects of scientific presentations, so is a whole chapter purely on the delivery aspects really necessary? We think it is. As we identified in Chapter 11, *Preparing scientific presentations*, presentations can be divided into two overlapping but nonetheless distinct areas: content, and the communication of the content (or 'substance' and 'style'). We said that both of these elements are important and you need to be careful not to focus on one at the expense of the other. Just because you have prepared a good presentation (the content) it doesn't mean that you will be able to deliver it in a competent fashion (the communication of the content). Communicating the content involves a very different set of skills to preparing the content, so this chapter will look at how you can communicate your content to best effect.

In this chapter we will begin by highlighting common concerns people have about presentations and how to address them, then we will think about how to use visual aids effectively, and finally some key techniques to help your presentations run smoothly. But first, let's think about some important differences between written and spoken presentations.

Written versus spoken presentations

A useful starting point when considering the delivery of a presentation that you have prepared is to think about the similarities and differences between written and spoken presentations. One of the reasons some students get anxious about giving spoken presentations is that they consider them to be very different from written presentations. Of course, in many respects the two forms are different, and we will come to the differences shortly, but there are many similarities too. In terms of the process that you need to go through in order to create written or spoken presentations, both forms require you to:

- analyse the question or brief;
- research the topic;

- plan and order the material;
- write the text or notes;
- review and re-draft the text or notes.

In addition to the above, in an academic context, both forms require you to present and analyse evidence, structure your argument, and reference your sources. When it comes to actually delivering the presentation however, the major difference between the two forms is that for a written presentation all that is required is for you to hand it in (whether that be on paper or via email or web submission), whereas delivering a spoken presentation requires a lot more effort.

There are other important differences between written and spoken presentations too. For example, when you are reading a text book or journal article and you come across a complex argument that you don't understand you can skip back and re-read the argument until you do understand it. Or if you come across a large quantity of data that is difficult to take in, you can study the data more closely, or even look up additional information in other sources, to make it easier to understand. When taking in information in the context of a spoken presentation, however, it is often much more difficult to assimilate complex arguments or large quantities of information simply because you do not have this ability to skip back as you would with written material. Equally, if, when you are reading, you find yourself getting tired and your attention is wavering, you can simply stop reading, have a break, and come back to it later when you are feeling more alert. Clearly, when listening to a presentation, apart from walking out and choosing to miss the rest of the presentation (which would be a bit rude), you don't have this option. So in a presentation, presenters have to work hard to make sure their listeners stay with them, both in terms of their understanding and their attention. This chapter will explain how you can do this.

12.1 Common concerns and how to address them

It is the differences between written and spoken presentations that are often the cause of concern or anxiety about delivering a presentation. For example:

- what if I'm so nervous I forget what I was going to say?
- what if the technology doesn't work?
- what if I lose my place in my notes?
- what if the audience don't seem to be following what I'm saying?
- what if I get asked questions that I don't know the answers to?

These, and other concerns, are important things to consider. Whilst listing possible concerns in this way can seem a little overwhelming, it is important to identify them so that something can be done about them. You might be able to relate to all of the above concerns, or perhaps to just a few of them, but it is likely that you will experience all of them at some point or other, so it will be useful to address each of them in turn.

12.1.1 Deal with anxiety

It's all very well saying 'deal with anxiety'; it is much more difficult to actually do it. It is also quite difficult to write about how to deal with anxiety, because different things make different people anxious. However, for the purposes of this chapter we are going to assume that standing up in front of a group of people to give a presentation is something that causes most people to be anxious or nervous. If it doesn't, it probably should!

We have deliberately titled this section 'deal with anxiety', rather than, for example 'get rid of anxiety' for a couple of reasons. Firstly, you can't get rid of all your anxiety so it would be unrealistic to try; and secondly, anxiety or nervousness can be helpful so, even if you could, you don't want to get rid of it altogether. Anxiety can motivate you to do the work needed so that you are effectively prepared for the presentation and it can also make you more alert and energized during the presentation. However, if your anxiety becomes too pronounced, difficulties can occur that may impair your ability to prepare effectively for and perform during the presentation.

A useful strategy in dealing with anxiety is to identify what your concerns are so that you can address them, which is what this section is about. In general terms, however, the following strategies can be helpful:

- try to replace negative thoughts with positive ones (see Table 12.1);
- practise steady, deep breathing before and during your presentation;
- have a bottle of water to hand – it will help stop your throat drying out, but better still, if your mind goes blank you can simply stop and take a sip, which usually gives you just enough time to gather your thoughts and regain your composure without the audience noticing.

TABLE 12.1 Replacing negative thoughts with positive ones.

Negative thought	Positive replacement
'I hate this!'	'This will only get better with practice.'
'It will be a disaster.'	'I will aim to do the best I can.'
'I never do any good at this kind of thing, it's bound to go horribly wrong.'	'Just because I had a problem with this is in the past does not mean that things are bound to go wrong.'
'I will fail my degree and never get the career of my choice if I don't do well in this presentation.'	'The marks for this presentation are only a small percentage of my overall degree. If I don't do as well as I would like there will be other opportunities to improve my marks.'

Try this: A one-minute presentation

Get together with a small group of friends. Each choose a topic with which you are comfortable and familiar, for example, your favourite hobby, holiday destination, or similar topic with which you are familiar. Now prepare a very short presentation on this topic. Each of you should now present the talk to the rest of the group. The talk should be no longer than one minute. One of the group should keep time.

12.1.2 Have a back-up plan

The second concern that we identified above was 'what if the technology doesn't work?' This is a legitimate concern and one that requires consideration. Broadly speaking there are two possible causes of technology apparently not working: first, it could be a genuine fault with the technology; or, secondly (and perhaps more commonly), it could be a lack of competence on the part of the person operating it. Both types of problem can be minimized by having a back-up plan, but be careful not to jump too quickly to using your back-up plan, when it could be, for instance, that you are just pressing the wrong button!

It is important, therefore, that you familiarize yourself with any technology that you intend to use. Amidst the pressure and stress of a presentation about to start, or already started, it can be easy to panic and not think straight. Make sure you understand how to use what you are using. If possible, practise not just with something like it, but with the *actual* technology in the *actual* room where you will *actually* be presenting. The closer you can get to practising with the real thing the more confident and prepared you will be when it comes to the presentation.

For those occasions when the problem is with the technology rather than the person operating it, this is where you will have to use your back-up plan. Some forms of technology are less reliable than others, and the more multi-faceted the output the more potential there is for it to go wrong. For instance, if you are using PowerPoint®, video, audio and a live internet connection – there are a lot of things that can potentially go wrong! You need to consider whether or not such a range of technology is necessary to your presentation. There would be much less risk if you used less technology and it would be much simpler. Indeed, it may even improve your presentation (more technology does not necessarily mean better presentations).

12.1.3 Make your notes work for you

The third concern that we identified above was 'what if I lose my place in my notes?' As we identified in Chapter 11, there is a lot you can do to help yourself here. Firstly, make sure you use an appropriate type of notes: outline or verbatim. Secondly, make sure you format them in a way that makes them easy to use and so it is less likely that you will either lose your place (a possible problem with verbatim notes) or forget what you were going to say (a possible problem with outline notes). It is a good idea to make it clear in your notes when you will start a new slide. Chapter 11 also highlighted the benefits of combining these note types to maximize the benefits, using outline notes where you don't need much prompting and verbatim notes where you do.

12.1.4 Help the audience follow what you are saying

The fourth concern was 'what if the audience don't seem to be following what I'm saying?' This is clearly an important consideration for the benefit of the audience, but it is also a significant factor in your own performance as a presenter. If you begin to notice, during your presentation, that your audience is distracted and perhaps seemingly confused, it can be very difficult to not let that influence your confidence and you can begin to lose your composure as a result. Obviously, it is important not to be over-sensitive here; it can be easy to interpret normal audience behaviour, such as someone yawning, as a signal that your presentation is going really badly, when this probably is not the case. It is likely that there will always be some people in the audience who have had a bad night's sleep, or are thinking about what they are going to do at the weekend, or perhaps just exhibiting body language and facial expressions that are out of sync with their genuine response. Don't let one or two negative responses unduly influence your confidence. Instead, there are several things you can do during your presentation to help your audience follow what you're saying:

- use verbal signposts;
- use your voice;
- be aware of your audience's needs.

Let's consider each of these in turn.

Use verbal signposts

In Chapter 11 we identified that one of the characteristics of effective presentations was that they should be well structured. If a presentation is well structured then the audience can see the direction the presentation is taking and so should be able to follow it more easily. We said that one of the ways you can do this is by letting the audience know what the structure will be; both at the beginning and as you go along.

For example, at the beginning of a presentation you should map out where you are going by telling the audience what you will cover. The example presentation brief that we introduced in Chapter 11 was:

> Prepare and deliver a five-minute presentation on high-temperature superconductors.

For this presentation, letting the audience know what the structure will be may involve you saying something like this (as seen in Chapter 11): 'This presentation is divided into three main sections. First, I'll give a brief explanation of what high-temperature superconductivity is. Then I'll describe some materials that exhibit superconductivity. Finally, I'll tell you about two specific applications: trains and the Hadron collider.'

In addition to mapping out the structure at the beginning of the presentation it is also helpful to give the audience pointers as you go along. If you were reading a written presentation you could simply flick through the pages and read the headings and subheadings, but clearly you can't do this with a spoken presentation. So help your audience follow what you are saying by telling them what the structure will be and then use pointers as you go along to make it clear where you are up to. The examples from Chapter 11 were:

- a single word such as 'First', 'Secondly', and so on;
- linking statements such as 'So that's the background to the technology, let's now think about its application…';
- concluding statements such as 'In conclusion…' and 'Finally…';
- finish with something like 'Thank you for listening'.

Use your voice

Use your voice to help your audience follow what you are saying. You can use it in many different ways by varying the volume, pace, and pitch.

Volume

Make sure that your voice is loud enough for your audience to hear clearly (it's surprising how many presenters don't do this – nervousness will often make you speak more quietly). Speaking too loudly or too quietly can make it difficult for your audience to follow your presentation. In normal conversation people tend to raise or lower their volume for emphasis. For example, a person may speak loudly when giving an instruction but softly when apologizing. Try to use these normal conversational variations in your own presentation.

Pace

Make sure that the speed of your delivery is easy to follow. If you speak too quickly (again, this is often caused by nervousness) or too slowly your audience will have difficulty following your talk. To add life to your presentation, try changing the pace of your delivery. A slightly faster section might convey enthusiasm. A slightly slower one might add emphasis or caution. If you are worried that you might speak to quickly then try to remember to breathe deeply and regularly as it will stop you rushing forward.

Pitch

The pitch of your voice also varies in day-to-day conversation and it is important to play on this when making a presentation. For example, your pitch will raise when asking a question; it will lower when you wish to sound severe.

Be aware of your audience's needs

If you are giving a five- or ten-minute presentation to a group of your peers and a tutor, then you will be limited as to how much you can actually do about your audience's needs. You will probably be on quite a tight schedule (often you will lose marks for running over time) and if there are several of you presenting, then it will often be a case of just making sure the presentations fit in the allotted session. Additionally, when you are relatively new to presenting, you will want to concentrate on just delivering what you have prepared rather than making any changes to your presentation as you go along in response to the reaction you are receiving. However, if you are at least half-way down a list of, say, six presentations that are scheduled for a session, there is one simple thing you can do (if your tutor doesn't suggest it): simply suggest a brief break. The break does not have to be long and it will probably be best if people don't even leave the room. However, if you suggest a break of a minute or so, while you are setting up, perhaps even suggesting that people stand up and walk around a bit, it should be just enough for people to wake up and re-energize themselves so that when you start your presentation they are that bit more attentive, making it easier for both you and them.

12.1.5 Respond to questions appropriately

The final concern we identified above was 'what if I get asked questions that I don't know the answers to?' This is a common concern because it is the bit of the presentation that you have least control over and it can be difficult to know what kind of questions you might be asked. If you are doing an assessed presentation, one important piece of information to find out is whether or not you are required to invite questions and if so whether or not your responses count towards the overall mark you receive. Whether or not the questions (or more accurately your answers!) count towards your mark or not there are a number of things you can do to make this part easier.

Create a list of potential questions and prepare answers

We said above that it can be difficult to know what kind of questions you might be asked, but it is not impossible. In fact, with just a bit of thought you can probably come up with several potential questions that you might get asked. Questions will usually focus on areas of the presentation that the audience found particularly interesting, or perhaps a section that needed more explanation or additional context. Obviously you can't predict all the questions you might be asked, but you will be able to predict some, and that will help to take some of the pressure off.

Say whether you will be taking questions and when

It is a good idea to state clearly at the beginning of your presentation whether you will be taking questions and, if so, when. In that way, the audience is clear about if and when they can ask questions but also it means that you are more in control of the situation, which is important: lack of control, as identified above, can make the questions part of a presentation difficult. Sometime people like to deal with questions as they go along as it can allow more interaction. However, it is usually easier to deal with questions at the end so you can deliver your talk free from interruptions, which you might find distracting.

Ask for clarification

When you get asked a question it is often unclear what the question actually means. If you are going to answer the question that the questioner wants to ask rather than the one you think they might be asking then it is important to ask for clarification. For instance, you might get asked the question:

Do all superconductors have the same structure?

A good way of clarifying the question is to rephrase it and then ask if that was what was meant. For example:

When you say 'all superconductors' do you mean you mean all Type 2 superconductors?

Clarifying questions in this way is useful for a number of reasons:

- it gives the questioner chance to refine the question;
- it ensures that the rest of the audience has heard and understood the question;
- it enables you to check that you have understood the question;
- it also gives you thinking time to prepare a better answer.

Answer the question

Preparing potential questions and answers, saying if and when you will take questions, and asking for clarification will all help you to answer questions more appropriately. However, when you actually open your mouth to answer the question make sure you:

- answer the question you have been asked rather than the one you want to answer (seeking clarification helps here);
- are focused and to the point – answer the question and then stop, don't start waffling;
- are honest about the limitations of your knowledge – say if you don't know the answer (it will usually be obvious if you are bluffing);
- say if a question is beyond the scope of the presentation ('That's a good question but my research for this presentation focused on examples and applications rather than the solid-state structures').

12.2 Use visual aids effectively

Using visual aids is relatively easy; it's using them appropriately that can be difficult. You will be limited to a certain extent by what is available, but equally important is to consider what is appropriate; just because you could use something doesn't mean that you should use something. The list in Section 12.2.1 shows the variety of visual aids available to you, and it is important that you recognize this variety; choosing the right form of visual aid is the first step in using them appropriately. Too often presenters default to using PowerPoint slides, without considering whether or not this is appropriate. For instance, a digital projector and laptop may be available to you to, but it may not be appropriate to give a PowerPoint presentation if you are simply required to give an informal update regarding a project to a small tutorial group. In this situation, a concise handout identifying key points that you can chat through might be more suitable.

12.2.1 Different types of visual aids

The more common types of visual aid are as follows:

- white board – large white board, usually fixed to the wall, can be written/drawn on with dry-wipe marker pens, either pre-prepared or spontaneously produced;
- flip chart – large pad of paper (A1), attached to a portable stand, can be written on with marker pens, either pre-prepared or spontaneously produced;
- handouts – usually A4 paper, handed out at an appropriate point (or points) during the presentation;
- objects – could be any physical object relevant to your presentation, shown from the front or handed round to your audience;
- video – shown on a television or via a laptop, various formats: VHS, DVD or digital media file;
- overhead projector or visualizer slides – hand written or printed onto transparencies, either pre-prepared or spontaneously produced;
- Microsoft PowerPoint slides (or equivalent) – produced on a computer and projected onto a screen via a digital projector, can be static or animated.

Each type of visual aid has advantages and disadvantages. These are summarized in Table 12.2.

12.2.2 When to use visual aids

Visual aids can be used throughout your presentation but too many visual aids can become distracting. Try to restrict the use of visual aids, therefore, to key parts of your presentation.

TABLE 12.2 Advantages and disadvantages of various visual aids.

	Advantages	Disadvantages	Suggestions
Whiteboard	• They are usually big – so lots of space to write/draw • They are particularly good for drawing equations because there is lots of space • Good for things you want to leave up and refer to at different points during your presentation • Good for recording comments or questions from the audience as you go along • Can be pre-prepared or spontaneous	• If you write on it during your presentation you will be turned away from the audience as you are writing, which limits communication • If you run out of space you will need to rub off at least some of what you've previously written • You need to be able to write legibly and large enough for the audience to see	• There aren't always appropriate pens available – so take your own • Write large and clear in dark colours (black or blue)
Flip chart	• Good for things you want to leave up and refer to at different points during your presentation • Good for recording comments or questions from the audience as you go along • Sheets can be torn off and stuck on the wall for reference – or flip back through the pad • Can be pre-prepared or spontaneous	• Usually smaller than a white board so not suitable for larger groups • If you write on it during your presentation you will be turned away from the audience as you are writing, which limits communication • You need to be able to write legibly and large enough for the audience to see	• There aren't always appropriate pens available or enough sheets – so take your own • Write large and clear • Try and keep to one topic per sheet
Handouts	• The audience has something to take away for reference • Particularly good for supplementary information, e.g. references	• If handed out at the wrong time they can be distracting because the audience stops listening to the presenter and starts reading the handout • If handouts are very detailed it is more likely that the audience will listen less	• Use them for information that is too detailed to put on a slide • Think carefully about when and how to give them out
Objects	• Not a commonly used form of visual aid so can have added impact • Other visual aids are mostly just looked at, objects give the audience something to	• If you have only one copy of the object it will take time to hand round (so maybe better to just show from the front)	• Make sure the object is directly relevant to the content of your presentation

TABLE 12.2 Cont'd

	Advantages	Disadvantages	Suggestions
	touch too, therefore more tangible and memorable • Molecular models are very effective aids.	• Because of the interest they can generate it is important that the object is central to your presentation, otherwise it will just be a distraction	
Video	• Can add interest and help grab attention	• Can be hard to find a directly relevant video • The technology can be a little unpredictable	• Make sure you keep the clip short • Check the technology works
Overhead projector or visualize slide	• Projected image is usually large enough for even large groups to see • Can be hand-written or printed • Can be good for showing the development of a process or building up a diagram • Can be pre-prepared or spontaneous	• You need to be able to write legibly and large enough for the audience to see	• Make sure text is large enough to be seen from the back of the room • Keep slides simple, clear and concise
PowerPoint slides	• Quick and easy to produce high-quality slides • Animation features can be helpful	• Some of the pre-set designs are unhelpfully formatted in terms of colour and background • Animation features can be very distracting if used inappropriately	• Keep slides simple, clear and concise • Use animation only when appropriate

Some suggestions of when to use visual aids are given below (we will address what visual aid to use shortly).

Introduction

You could use visual aids during the introduction to do some of the following:

- display the title of your presentation and perhaps your name;
- define particular terms or units you are going to use that your audience might not be familiar with;
- indicate what the structure of your presentation will be;
- display an image that introduces your subject or theme;

- highlight a question that you intend to answer during the course of your presentation.

Main points

You could use visual aids during the main body of your presentation to do some of the following:

- highlight your main points as they arise, using appropriate words or images;
- indicate the transition between main points;
- summarize your results or data visually using graphs, tables, reaction schemes, etc.;
- display key evidence from your research to support your argument.

Conclusion

You could use visual aids during the conclusion to do some of the following:

- summarize your main points;
- present your conclusion in a succinct phrase or image;
- list your references to enable your audience to read more on the topic;
- give your audience something to focus on other than you!

12.2.3 What visual aids to use

In Chapter 11 we suggested a number of questions to help you select supporting information for a presentation. You can use these same questions to help you decide whether or not to use a particular visual aid. The fundamental issue is whether the material represented in the visual aid is relevant to the presentation, and if it is you then need to ask yourself; will this visual aid make my presentation:

- clearer?
- more authoritative?
- more interesting?

And if your visual aid doesn't do any of the above three things, don't use it.

Visual aid choice exercise

For our sample presentation on high-temperature superconductors a possible outline for the presentation was: 'Short explanation of what high-temperature superconductivity is, materials that exhibit high-temperature superconductivity, and examples of applications.' Imagine you were considering using the following visual aids for the above presentation:

- a flip chart page outlining the structure of your presentation (Figure 12.1);
- a handout giving details of compounds that exhibit superconductivity;
- a molecular model of a perovskite structure;
- a video clip (4 minutes 33 seconds) of a superconducting levitating model train from YouTube (http://www.youtube.com/watch?v=TeS_U9qFg7Y);
- a hand-drawn overhead projector slide or visualisor sheet showing the synthetic strategy used to prepare YBCO superconductors;
- an image of a YBCO structure on a PowerPoint slide (Figure 12.2).

FIGURE 12.1 Flip chart summarizing presentation content.

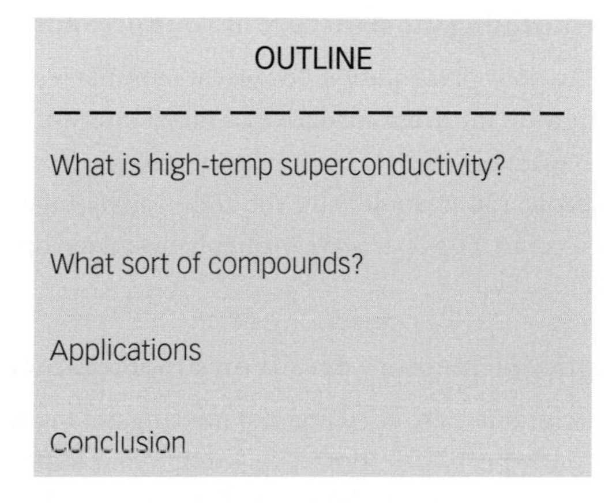

FIGURE 12.2 The structure of the YBCO superconductor.

Source: Atkins, P., Overton, T., Rourke, J., Weller, M. and Armstrong, F., 2006 *Inorganic Chemistry*, 4th edn, Oxford: Oxford University Press.

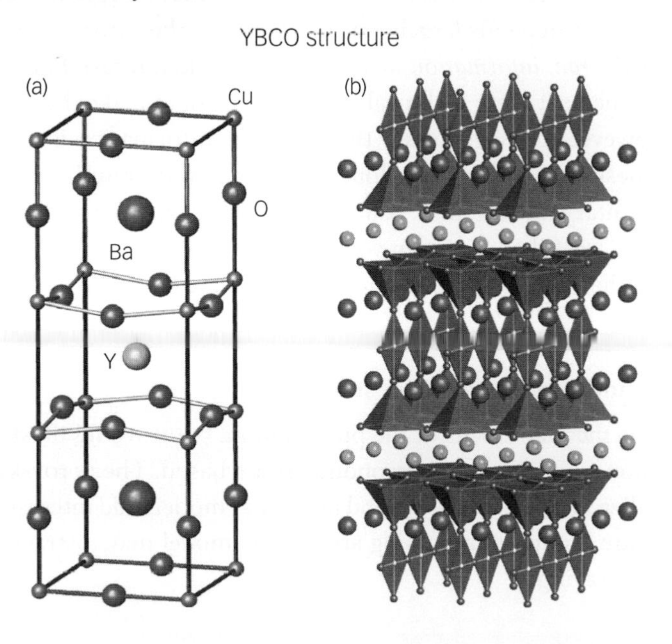

Based on the questions listed above, decide whether or not you should include these visual aids in your presentation, and why.

Visual aid choice exercise feedback

We have identified that the fundamental issue when choosing whether or not to use a visual aid is its relevance: is the material represented in the visual aid relevant to the presentation? If you decide it *is* relevant you then need to question whether the visual aid makes your presentation clearer or more authoritative or more interesting. We will deal with each of the suggested visual aids in turn, testing them against each of these criteria.

- **A flip-chart page outlining the structure of your presentation**

This is clearly relevant to your presentation because it summarizes its content. Whilst it won't make the presentation more authoritative or more interesting, it could help make it clearer. If you were to use a flip chart in conjunction with another form of visual aid, for example an overhead projector, it would have the added advantage of the outline always being there for you to refer to. Flip charts are probably best used only when the audience is relatively small and close by.

- **A handout giving supplementary details on superconducting materials**

This is potentially relevant but there is a danger of missing the main focus of the presentation. The presentation is primarily about explaining what superconductivity is with some examples of compounds and applications. Whilst brief discussion of the different compounds that exhibit superconductivity may be helpful too much detail could be distracting or add unnecessary complexity. On the other hand, it would provide something useful for the audience to take away with them. It may add greater authority to the presentation, but this depends largely on the source of the information (see Chapter 5, *Working with different information sources*). It could add interest, but this depends on how you communicate it (very technical information could make the presentation less interesting). However, if you genuinely think that the information is useful, a handout is probably the best format for it. Handouts are good for communicating (among other things) supplementary information; your audience can just refer to the handout if they are interested. Be careful about when you give them out though: you don't want your audience having their heads down reading when you are trying to move on to the next point!

- **Molecular model of a perovskite structure**

This shouldn't be the main focus of the presentation, but it would illustrate effectively the structure upon which many superconductors are based. The perovskite structure is difficult to visualize in two dimensions and molecular models add interest and aid understanding. Be aware though that passing around the model may distract your audience from what you are saying.

- **A video clip (4 minutes 33 seconds) of a superconducting levitating model train from YouTube (http://www.youtube.com/watch?v=TeS_U9qFg7Y)**

This might sound exciting and it is. But you have to think whether it is appropriate. It certainly illustrates levitation brilliantly and is directly relevant to your levitating-train application. It is potentially relevant in terms of the brief because it is about one of the applications you discuss, but it's probably less relevant in terms of the level. In Chapter 11, we said that a presentation should not only be relevant with regard to the brief, it should also be relevant with regard to the audience. A video designed for a popular audience is probably not going to be relevant for academic purposes. It's unlikely to add authority because, even though it is from a credible source, the level, as identified above, is too popular. It might add interest but the clip is far too long; if you were to use the entire clip it would take up more than half your allotted time for your presentation.

- **A hand-drawn overhead projector slide or sheet for a visualizer showing the synthetic strategy used to prepare YBCO superconductors**

If you like the idea of using cutting-edge technology you might be reluctant to use a hand-drawn overhead projector slide. But many lecture theatres are now equipped with digital visualizers that do the same job but project through a digital projector. However, let's think about our criteria. Such a slide would be irrelevant because the original brief did not include synthetic strategies. It may add authority and interest, however, because it focuses closely on practical chemistry rather than applications.

- **An image of a YBCO structure on a PowerPoint® slide (Figure 12.2)**

The image is taken from a reputable source and so adds authority. It is relevant as this is the structure of the best-known superconducting materials. It is sensible to use a professional-looking picture like this for such a complex structure as it would be very difficult to either hand draw or to build a model of. It is appropriate to use images such as this but make sure you have a clear purpose for using them (and as long as you acknowledge the source, see Chapter 10, *Avoiding plagiarism*). Don't just think, 'there's a picture of YBCO structure – I'll use that', ask yourself why you need to use it.

12.3 Master some key techniques

We have addressed some common concerns that people often have about giving presentations and considered how to use visual aids appropriately. Finally, it's important to think about some key techniques that you need to develop if you are going to deliver your presentation effectively.

12.3.1 Relax!

The first key technique is to try to relax. Clearly, this is easy to say but much more difficult to actually do. However, if you can relax, even just a little bit, it will help your presentation a great deal. A presenter who is tense will deliver their presentation in a stilted, unnatural, and nervous manner, whereas a presenter who is relaxed (or slightly less tense at least) will be more natural and more confident in their delivery. We addressed many of the issues that will help you to be more relaxed in Section 12.1.1, *Deal with anxiety*, but we wanted to highlight it here as a key technique as well because it makes such a positive difference to a presentation.

12.3.2 **Be conversational**

In Chapter 11 we highlighted the importance of getting to the stage where you have practised your presentation sufficiently so that your notes (whatever their format) act as a prompt rather than a script, thus freeing you up to think about the communication, as well as content issues. Something to aim for is to try and be conversational. Being conversational doesn't necessarily mean being informal or chatty (this is often not appropriate in an academic context) but it does mean that you are speaking in a natural way, as if you were actually having a conversation. Clearly, a five-minute presentation is a fairly one-way conversation (!) but if you can aim to express yourself in normal, everyday (though still appropriate) language it will help the presentation to be more engaging, and being engaging was one of the key characteristics of effective presentations that we mentioned in Chapter 11.

12.3.3 **Think about your body language**

Body language is an important and often neglected element of a presentation. It includes facial expressions and eye contact as well as what you do with your hands and where and how you situate yourself in the room. Body language can be both positive and negative: it can help a presentation be more engaging, or it can distract from and even counteract the presentation's content. If you have practised your presentation in front of your friends, a mirror, or a digital camera (as suggested in Chapter 11) you will be more aware of how your body language influences the way you present. Whilst it can be counterproductive to become too self-conscious of your body language, here are some suggestions to make sure you use your body language positively.

Think about how you will use the available space

Where are you going to stand in the room (you might prefer to sit but it's more normal to stand and it helps you project your voice better)? Where will you put your notes? If you are using PowerPoint slides or an overhead projector you need to be able to stand somewhere that doesn't obscure anyone's view of the screen but still allows you to be close enough to change slides without getting in anyone's way or tripping over cables. Spend some time thinking about this before you start presenting because once you have started your presentation you want to be able to concentrate on what you are saying rather than thinking about where the best place is for you to stand.

Think about what you will do with your hands

People use their arms and hands in every day conversation to add emphasis or to help describe events. Presenters will therefore look rather awkward if they keep their hands in their pockets or rooted firmly at their sides. Equally, you don't want your hands to move in a way that distracts your audience's attention such as nervous, repetitive movements. Instead, use your hands to emphasize and enhance what you want to say verbally. As long as your use of hands is controlled and purposeful, then it has a key role to play in supporting your verbal communication. If you can try to relax a little and be conversational then it is more likely that what you do with your hands will be more natural and more in sync with what you are saying.

Make eye contact

Often during an assessed presentation the presenter only makes eye contact with the person doing the assessing, thereby excluding everybody else. If you were to have a conversation with someone you would make eye contact with them, and not doing so would probably be considered a bit rude. So when presenting, try and make eye contact with everyone. A common error for nervous students is for them to focus solely on the tutor. This excludes the rest of the audience and can be uncomfortable for the tutor. It is likely that you will only be presenting to relatively small number of people, so making eye contact with everyone is a realistic aim. If, however, you are presenting to larger groups just try and make eye contact with all parts of the room. Making eye contact is difficult if you are reading your notes, so make sure follow the guidelines on how to use verbatim or outline notes in Chapter 11.

Smile

If you are nervous it can be difficult to smile, but smiling can make a big difference to how people react to you. Clearly, you don't want to have an inane grin on your face for the entire presentation, that would be unnatural and look ridiculous, but making a point of trying to smile, at least before you start, will help your audience feel more at ease. It is likely that at least someone in the audience might smile back at you, which in turn helps you feel more at ease too.

12.3.4 Show some enthusiasm!

Finally, show some enthusiasm! If you appear bored by and uninterested in what you are saying your audience will probably react in the same manner. Conversely, if you can show genuine interest and enthusiasm it is more likely that your audience will be willing to listen attentively and work at following what you are saying.

* Chapter summary

There are similarities between written and spoken presentations, but there are significant differences too. It is often the differences that are the cause of concern or anxiety associated with delivering a presentation. This chapter has addressed these common concerns and also focused on how to use a range of visual aids appropriately and the importance of mastering some key techniques. Following these guidelines will ensure that you deliver your scientific presentations much more effectively. Some key tricks to remember:

- try to think positive;
- be prepared for technology failures;
- use the best type of notes for your style;
- use verbal signposts to direct your audience through your talk;
- watch the pace, volume and pitch of your voice;
- remember to breathe. It helps your pace, calms your nerves, and gives you thinking time;
- choose appropriate visual aids;
- make eye contact with the audience;
- try to be enthusiastic;
- practice, practice, practice!

Chapter 13

Creating academic posters

Introduction

Poster presentations are a common academic format and are a regular feature at academic conferences. Whilst, as an undergraduate, it is unlikely that you will be presenting a poster at a conference, the poster format is sometimes used as a form of assessment for undergraduate courses. This chapter will explain the purpose of academic posters and suggest seven key steps for creating them effectively.

13.1 The context of poster presentations

If posters are used as a form of assessment at undergraduate level then they have a different purpose from that of an academic conference. However, it is helpful to understand the context from which they come in order to know what they are supposed to achieve.

Academic conferences are important opportunities for researchers to present their work, allowing them to:

- publicize their findings;
- promote discussion;
- create opportunities for collaboration with other researchers.

Clearly, not everyone at a conference will be able to present their research orally, so poster presentations are often used to allow lots of people to present their research at once. Typically, posters are size A1 (eight times bigger than A4) or A0 (16 times bigger than A4) and are attached to display boards arranged throughout an open-plan room or conference hall. Time will be allocated in the conference schedule for delegates to view the posters, usually with opportunities to discuss the work with the presenter. A poster presentation is more limited than an oral presentation in terms of the amount of information that can be communicated, therefore posters should be designed to summarize the important elements and promote discussion.

Depending on how large the conference is and how well it is organized, delegates may have brief details of posters on display at the conference in advance, including the title of the poster and possibly an abstract. If these details are available then delegates might scan the list of posters on display to identify those they wish to view. Alternatively (or additionally), delegates may simply browse the posters themselves and view those that catch their eye. Either way, the posters that are viewed the most will be those that stand out from the crowd in some manner.

13.2 Seven key steps to creating your poster

Now that we have explained the context of academic posters we need to identify how to create them effectively. There are seven key steps involved in the process; the first two relate to the content of the poster, the next four are about design issues, and the final one is about ensuring that it is free from errors. This may seem a little unbalanced but we have dealt a lot already in this book with writing for chemical science, and it is usually the design aspects, rather than the content aspects, where students need the most guidance.

The seven steps, therefore, are:

1. get the academic content right;
2. cut the text down to between 300 and 500 words;
3. format the type;
4. prepare your images;
5. design the layout;
6. choose a colour scheme;
7. check it very carefully.

13.2.1 Get the academic content right

First, you need to make sure that you get the academic content right. This is the foundation for everything else that follows. If you don't do this first you will waste a lot of time formatting and laying out content that you end up not using. Usually, for undergraduate assessment purposes, you will be given a specific brief for your poster (often based on another aspect of assessment such as a practical report, essay topic or research method). This brief will determine the parameters of the academic content. In addition to the academic content you may also be given guidance on the audience the poster is intended to be designed for. The intended audience will determine the level of the academic content that is appropriate.

Specialist

If the audience is from the same specialist field as the presenter, the presenter can assume that the people viewing the poster will have a high level of knowledge about the subject. It is therefore acceptable for the presenter to use technical language and terms on the poster.

Related

If the audience is from a related field to the presenter, the presenter can only assume that the people viewing the poster will be familiar with the generalities of the discipline. The presenter, therefore, needs to be careful about using technical language and terms and should avoid it where possible.

General

If the audience is from the general public the presenter can only assume general knowledge. The people viewing your poster will probably have no familiarity with the discipline or subject area. The presenter, therefore, needs to avoid using technical language and terms and should use only basic descriptions.

For the purposes of this chapter we will use the following brief for a poster.

> Design and produce an academic poster summarizing an application of an instrumental analytical technique in chemical sciences. The intended audience is first-year undergraduate chemical science students. The poster needs to be size A1 and landscape orientation.

The application we will use is *analysis of arsenic in groundwater*. We will follow this example through in the remaining six steps of the process.

13.2.2 Cut the text down to 300–500 words

The second step in creating your poster is to use an appropriate number of words. Clearly, from the title of this section we are assuming that the problem you will have is having too many words rather than too few. That's because a poster usually requires you to take a piece of writing you have already created for another purpose and summarize its content. At an academic conference the original writing will usually be some form of research, at an undergraduate level, as identified above, it will usually be based on another aspect of assessment such as a practical report, essay topic or research method. The recommended number of words for a poster is between 300 and 500; this may not seem like much (and it isn't!) but it is important to keep the number of words this low in order [illegible] an a poster that is both readable from a distance and conveys key information quickly. The number of words affects readability because the greater the number of words the smaller the font size needs to be in order to accommodate them on the page, and so the less readable it becomes. Also, if the poster contains too much text, the amount of information can be overwhelming to the reader, making it less likely that passing delegates will spend time viewing it. To get an idea of what 300 and 500 words looks like see Figure 13.1 and Figure 13.2, respectively.

FIGURE 13.1 A poster with 300 words.

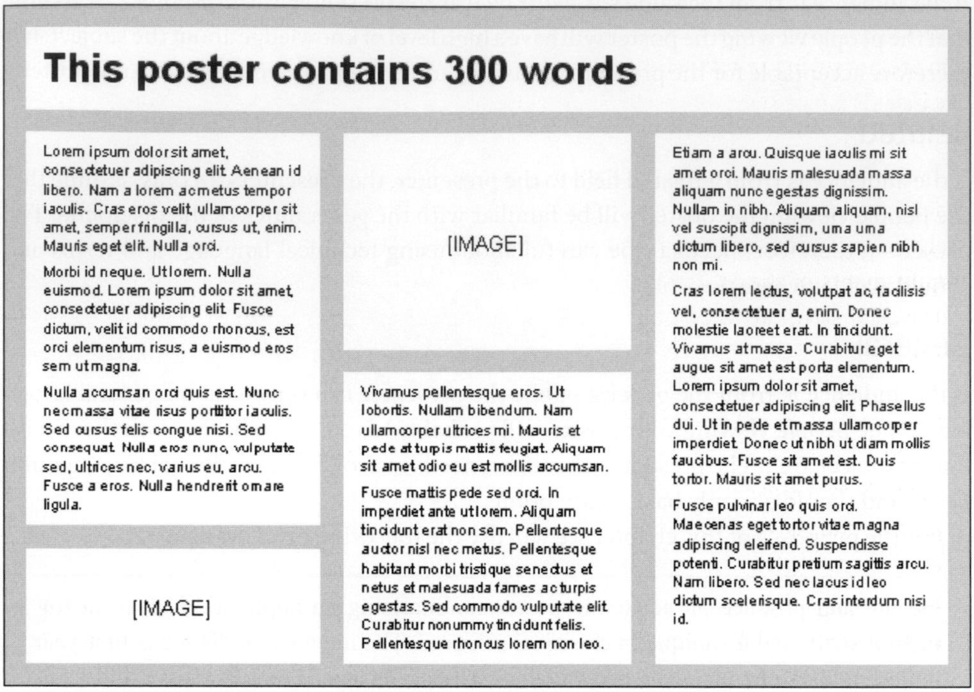

FIGURE 13.2 A poster with 500 words.

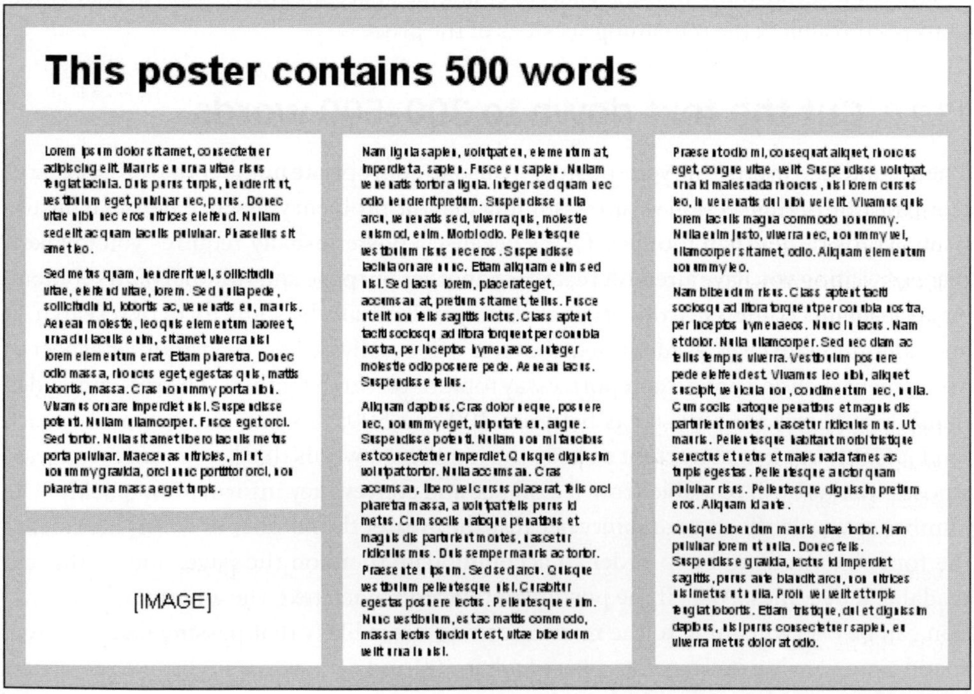

When reducing the text down to between 300 and 500 words you need to remember that you are trying to make it easy for a person who is not familiar with the content to understand it quickly. It can be helpful to try and think of it a bit like a trailer for a movie: you are not trying to tell people everything there is to know about the subject, rather you are just giving an overview in order to summarize the main points and encourage people to find out more.

For our poster on analysis of arsenic in groundwater we have summarized the current research techniques that are involved in the process and tried to explain them appropriately for an audience of first-year undergraduate biological sciences students, as per the requirements of the brief. We have used a conventional scientific report pattern to provide a clear structure and endeavoured to focus only on the essentials of arsenic analysis to create a succinct text. The words for our arsenic analysis poster are as follows.

Arsenic is a metalloid that can be found in the environment in rocks, soil, and water. It occurs naturally as the minerals arsenopyrite (FeAsS), realgar (AsS), and orpiment, (As_2S_3). Levels of arsenic in the soil can be found to be from 0.1 to 40 ppm. Arsenic can enter groundwater by erosion, dissolution, and weathering. Other sources include geothermal springs and volcanic activity. It can also be released by brines that are used to produce natural oil and gas, as a by-product of gold and lead mining and the combustion of coal. Industrial contaminations occurs through production of lead-acid batteries, paper production, glass, and cement manufacturing. Arsenic occurs in drinking water as As(III) or As (V)). Organic arsenic species are concentrated in seafood but are much less harmful to health than inorganic arsenic compounds as they are readily eliminated by the body. The European Union and the World Health Organisation recommend arsenic limits in drinking water of 0.01 ppm. Arsenic poisoning or arsenicosis results in many ailments such as skin cancer, cancers of the bladder, kidney and lung, gangrene, diabetes, high blood pressure, and reproductive disorders. It is obviously important to be able to detect low levels of arsenic in aqueous samples.

Groundwater contains a variety of inorganic and organic arsenic species and these must be separated by chromatography before determination. Arsenic can be determined by a variety of analytical methods including colorimetry, atomic absorption spectrometry, XRF, ICP, and ICPMS. In this project the arsenic species were separated by ion exchange chromatography and measured by ICPMS.

Common arsenic species in groundwater are shown in Table 1.

A Dionex ion-exchange column was used with gradient elution with 2 mM tetramethylammonium hydroxide and 10 mM ammonium carbonate binary mobile phase. Six arsenic species were separated over a period of 20 min in the order AsB, DMA(V), MMA(V), As(III), and As(V).

A Shimadzu inductively coupled plasma mass spectrometer with a miniaturized torch was used for the detection. The eluent from the ion-exchange column was introduced into the plasma by a nebulizer. Standards of each of the arsenic compounds were prepared in water over the range 0–50 ppm. Six-point calibration curves for the arsenic species were obtained by plotting the peak areas against the concentration of arsenic for each species. The calibration

TABLE 1 Common arsenic species in groundwater.

Arsenic species	abbreviation	formula
Arsenite	As(III)	$As(OH)_3$
Arsenate	As(V)	$AsO(OH)_3$
Monomethylarsonic acid	MMA(V)	$CH_3AsO(OH)_2$
Dimethylarsinic acid	DMA(V)	$(CH_3)_2AsOH$
Arsenobetaine	AsB	$(CH_3)_3As^+Ch_2CH_2COO^-$

curves gave coefficient coefficients in excess of 0.999. The detection limits ranged from 0.071 to 0.40 ppm. The concentrations of each of the arsenic species in the groundwater sample are given in Table 2.

Conclusions

TABLE 2 Arsenic concentrations.

Arsenic species	Concentration/ppm
As(III)	5
As(V)	11
MMA(V)	0
DMA(V)	0
AsB	0

An effective method for determination of arsenic in groundwater was developed. The levels of organoarsenic compounds were below the recommended limit. However, there was significant inorganic arsenic present.

B.K. Mandal, Y. Ogra and K.T. Suzuki, Identification of dimethylarsinous and monomethylarsonous acids in human urine of the arsenic-affected areas in West Bengal, India, *Chen. Res. Toxicol*., 2001, **14**, 371–378.

Z. Gong, X. Lu, M. Ma, C. Watt and X. Le, Arsenic speciation analysis, *Talanta,* 2002, **58**, 1, 77–96.

T. Nakazato, T. Taniguchi, H. Tao, M. Tominaga and A. Miyazaki, Ion-exclusion chromatography combined with ICP-MS and hydride generation-ICP-MS for the determination of arsenic species in biological matrices, *J. Anal. At. Spectrom.*, 2000, **15**, 1546–1552.

The above text contains 486 words (not including the references). The structure follows a conventional scientific report pattern: introduction, methods, results, conclusion, and references.

13.2.3 **Format the type**

Thirdly, you need to format the type. When formatting type for a poster, you need to remember two important principles: to use consistent styles, and to group sections of text appropriately.

Use consistent styles

Using consistent styles helps achieve what graphic designers call this the principle of **resemblance**. It means that text that performs the same function should always be formatted in exactly the same way. For instance, if you decide that the subheadings should be Arial, 10 point, bold, left-hand justified with an after paragraph line spacing of 0.2 lines then all your subheadings should be formatted in exactly this way. Similarly, you might decide that the captions for figures should be Arial, 6 point, regular, right-hand justified with a before and after paragraph line spacing of 0.2 lines, in which case all your captions should be formatted in exactly this way.[1]

There are a couple of reasons for using consistent styles: firstly, it makes the poster look neater; secondly it makes the structure of the poster easier to understand because you know which text is performing what function. An example of using consistent styles is shown in Figure 13.3.

Group sections of text appropriately

Grouping sections of text appropriately helps achieve what graphic designers call the principle of **proximity**. It means that text that belongs together is grouped together; this is usually achieved using paragraph spacing. Figure 13.5 illustrates inappropriate spacing of text; there is a whole line space between each section and in the bulleted list. The result is that the headings and bullets float around and it is not immediately clear which piece of text belongs to which section.

Figure 13.6 illustrates more appropriate spacing of text; paragraph spacing has been used to make it clear which pieces of text the headings and bullets belong to.

Line spacing, alignment, and text formatting

In addition to thinking about the principles of resemblance and proximity when formatting the type, it is also important to consider line spacing, alignment, and text formatting. Figure 13.7 illustrates the effect of line spacing on body text; body text (which is usually

---

1 Note that 10 point and 6 point are small font sizes, but posters are often created in A4 then enlarged to A1 or A0, making the finished text eight or 16 times bigger respectively. See Box 13.1.

FIGURE 13.3 Using a limited number of consistent styles.

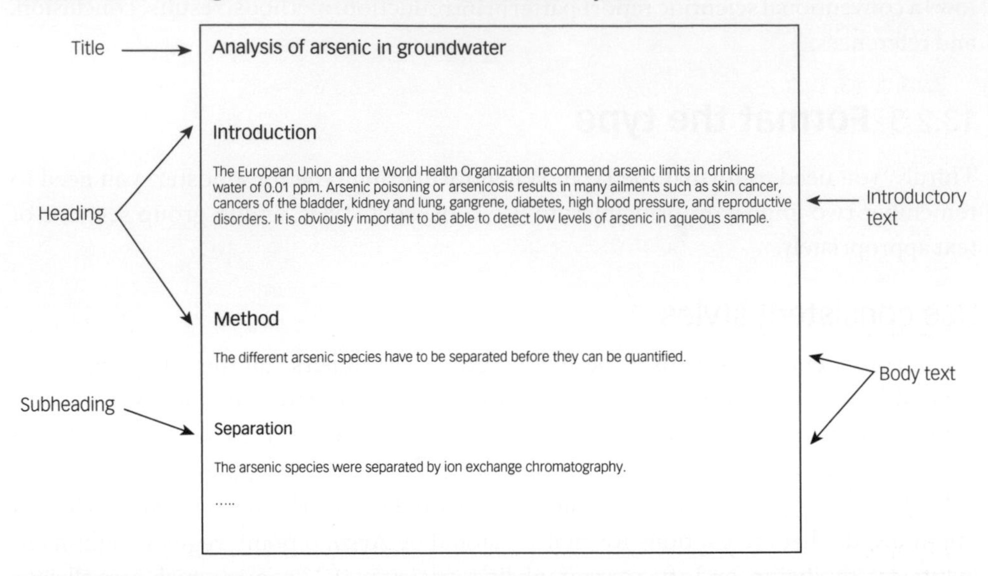

BOX 13.1 How do I know what font size to use?

Posters are often created at A4 size and then enlarged to the appropriate size for a poster, usually either A1 or A0. This can make it difficult to know what font size is appropriate to use because you are not creating your poster at the size of the finished product. The magnification factor from A4 to A1 is approximately 300 per cent (it is actually 283 per cent, for reasons that we won't go into here) and from A4 to A0 is 400 per cent. This relationship is illustrated for you in Figure 13.4.

FIGURE 13.4 A4 paper to A1 or A0.

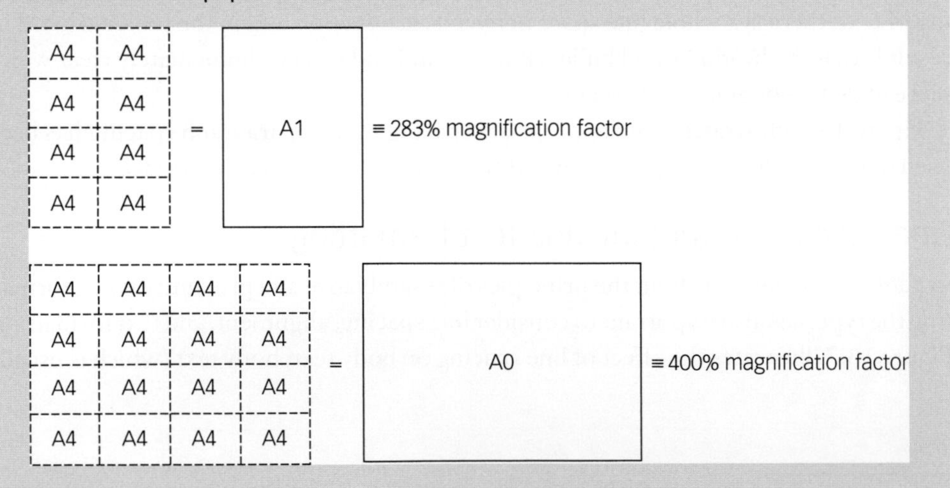

Therefore, a font size of 10 point at A4 is equivalent to a font size of 40 point at A0. Approximate appropriate font sizes to use for A1 and A0 posters are shown in Table 13.1.

BOX 13.1 cont'd

TABLE 13.1 Suggested font sizes to allow appropriate magnification.

Style	A4 to A1	A4 to A0
Title	**18 point Arial Bold**	**16 point Arial Bold**
Heading	**13 point Arial Bold**	**12 point Arial Bold**
Subheading	**11 point Arial Bold**	**10 point Arial Bold**
Inrtoductory text	**11 point Arial Bold**	10 point Arial regular
Body text	**10 point Arial Bold**	8 point Arial regular
Caption	**8 point Arial Bold**	6 point Arial regular

FIGURE 13.5 Inappropriate proximity of text.

Method

Groundwater contains a variety of inorganic and organic arsenic species and these must be separated by chromatography before determination. Methods of determination include:

colorimetry

atomic absorption spectrometry

XRF

ICP

ICPMS.

Separation

A Dionex ion-exchange column was used with gradient elution with 2 mM tetramethylammonium hydroxide and 10 mM ammonium carbonate binary mobile phase

FIGURE 13.6 Appropriate proximity of text.

Method

Groundwater contains a variety of inorganic and organic arsenic species and these must be separated by chromatography before determination. Methods of determination include:

- colorimetry
- atomic absorption spectrometry
- XRF
- ICP
- ICPMS.

Separation

A Dionex ion-exchange column was used with gradient elution with 2 mM tetramethylammonium hydroxide and 10 mM ammonium carbonate binary mobile phase

the smallest text on the page other than captions) is much easier to read if the line spacing is slightly increased.

Titles that run to more than one line, however, often benefit from the reverse: compressed line spacing, as shown in Figure 13.8. However, where possible, it is better if you can just make your titles more succinct so that they can fit on one line.

FIGURE 13.7 Line spacing and body text.

Line spacing = 1.0
Arsenic can enter groundwater by erosion, dissolution, and weathering. Other sources include geothermal springs and volcanic activity.

Line spacing = 1.5
Arsenic can enter groundwater by erosion, dissolution, and weathering. Other sources include geothermal springs and volcanic activity.

Line spacing = 2.0
Arsenic can enter groundwater by erosion, dissolution, and weathering. Other sources include geothermal springs and volcanic activity.

FIGURE 13.8 Line spacing and titles.

**Large text (e.g. titles)
probably needs decreased
line spacing . . .**

Line spacing = 0.9

**. . . because it looks a bit
strange with increased line
spacing.**

Line spacing = 1.1

It is recommended that you left-hand justify all body text (as shown in Figure 13.9) because fully justifying the body text results in awkward gaps on some lines making the text more difficult to read (as shown in Figure 13.10).

Finally, on formatting the type:

- use the same font throughout your poster (unless using a complementary font for headings);
- set the headings in **bold;**
- use *italics*, underlining and CAPITALS sparingly (only when convention dictates);
- break up any large areas of text with subheadings.

FIGURE 13.9 Left-hand justified body text.

Arsenic is a metalloid that can be found in the environment in rocks, soil, and water. It occurs naturally as the minerals arsenopyrite (FeAsS), realgar (AsS), and orpiment, (As_2S_3). Levels of arsenic in the soil can be found to be from 0.1 to 40 ppm. Arsenic can enter groundwater by erosion, dissolution, and weathering. Other sources include geothermal springs and volcanic activity.

FIGURE 13.10 Fully justified body text.

Arsenic is a metalloid that can be found in the environment in rocks, soil, and water. It occurs naturally as the minerals arsenopyrite (FeAsS), realgar (AsS), and orpiment, (As_2S_3). Levels of arsenic in the soil can be found to be from 0.1 to 40 ppm. Arsenic can enter groundwater by erosion, dissolution, and weathering. Other sources include geothermal springs and volcanic activity.

13.2.4 **Prepare your images**

The fourth step is to prepare your images. Images can be photographs, diagrams, graphs, molecular structures, reaction schemes, or tables. The appropriate use of images can benefit a poster considerably, not just in terms of its visual impact, but also in terms of how readily understandable the content of the poster is. We will deal with each of these image types in turn.

Photographs

There are three key points you need to remember about any photographs you use on a poster.

- They need to be relevant to the content of your poster, not just put in to brighten it up.
- If they are not your own photographs they need to be appropriately referenced (in the same way as you would reference text that wasn't your own).
- They need to be of an appropriate resolution.

The resolution is especially important if you are creating a poster in A4 format and then enlarging it to A1 or A0. In order for photographs in particular, and images generally, not to degenerate, they need to be of sufficient quality to start with – that is, they need to be of a sufficiently high resolution. The resolution of digital images is measured

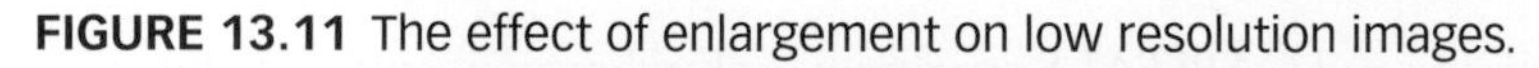

FIGURE 13.11 The effect of enlargement on low resolution images.

in pixels per square inch. Photographs copied from the internet usually have a resolution of 72 pixels per inch and so will look 'grainy' or 'pixelated' when enlarged, as shown in Figure 13.11.

For a printed poster, choose images that have a resolution of at least 300 pixels per inch at the size at which they will appear on the poster. In all likelihood you will want to enlarge the image when you reproduce it for the poster: and as soon as you enlarge it, the resolution will decrease. For example, if you have an image that is two centimetres square and has a resolution of 300 dpi, when you enlarge it to become four centimetres square, its resolution will drop to 150 dpi (which is too low a resolution for printing purposes). Instead, start with an image that has a resolution that's well above 300 dpi. Then, when you enlarge it, the resolution will drop – but hopefully not below 300 dpi.

Diagrams

A diagram can be a very useful way of communicating information in a quick and succinct manner. Therefore, they are of particular use in posters, where communicating information in a readily understandable and concise fashion is very important. A well-chosen diagram can save a lot of words or make text descriptions much easier to understand. In our example poster, Figure 13.12 supplements the information in the first paragraph of the methods section to make it more readily understandable.

Graphs

When making graphs for posters, you need to think carefully about how you format them as viewing them from a distance means that default formatting is usually inappropriate. The graph in Figure 13.13 is difficult to view from a distance because of the relatively small font, the shaded plot area and the excessive use of horizontal lines. All these formatting features are default features in Microsoft Excel®. As a general rule, when it comes to the formatting of a graph, don't accept the default formatting provided by your software.

FIGURE 13.12 A schematic representation of ICPMS.

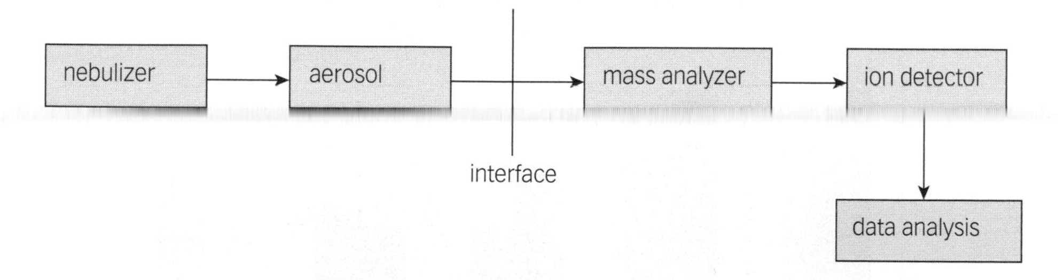

FIGURE 13.13 An inappropriately formatted graph.

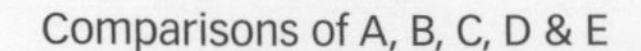

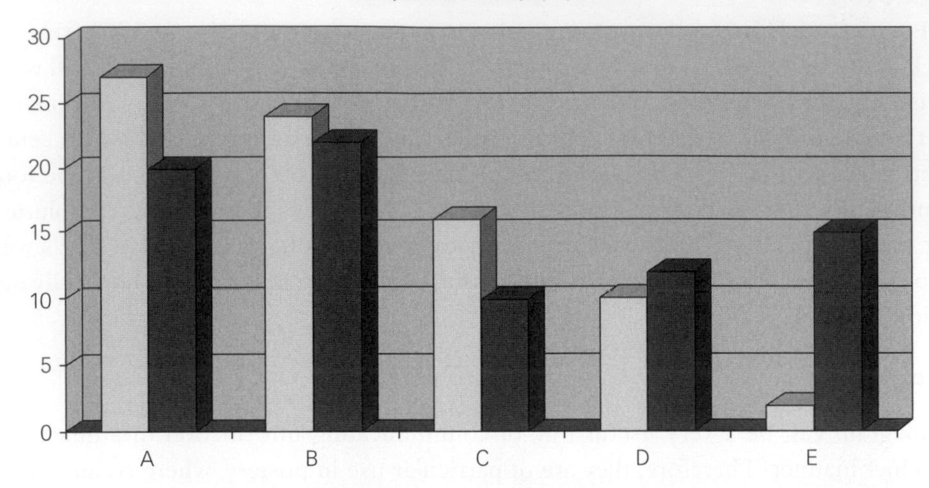

Figure 13.14 shows a more appropriately formatted graph. The font size has been increased, the shading on the plot area has been removed, and the use of horizontal lines has been minimized. Additionally, the colours of the bars have been altered to provide greater contrast; making the two sets of figures easier to distinguish between. Again, as with photographs and diagrams, if the data is not your own you need to reference it appropriately.

Tables

Finally, in this section on preparing images, let's consider tables. The principles that applied to graphs apply equally to tables: format them in such a way so as to make them clear, simple and easy to see. Again, default formatting is usually inappropriate as it tends to make the text too small and the grid lines can make patterns in the data difficult to discern. As a general rule, don't format tables that look like spreadsheets, as shown in

FIGURE 13.14 A more appropriately formatted graph.

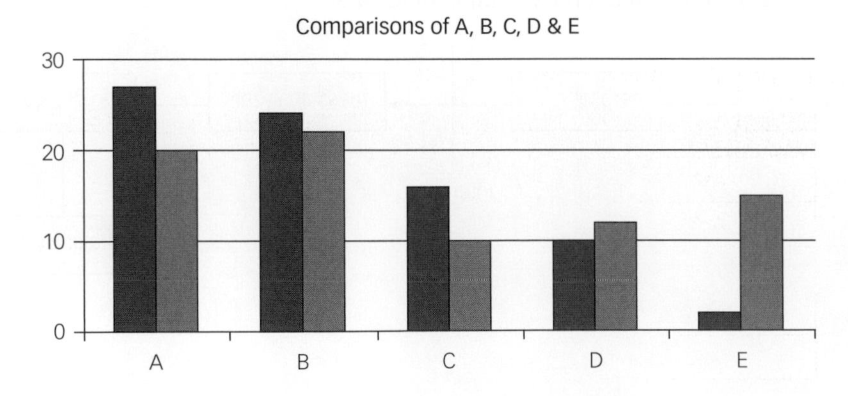

FIGURE 13.15 Default formatted table.

Arsenic species	Formula	Concentration/ppm
As(III)	$As(OH)_3$	5
As(V)	$AsO(OH)_3$	11
MMA(V)	$CH_3AsO(OH)_2$	0
DMA(V)	$(CH_3)_2AsOH$	0
AsB	$(CH_3)_3As^+Ch_2CH_2COO^-$	0

Figure 13.15, as they are more difficult to interpret, even when you are dealing with a small data set.

Figure 13.16 shows a table that has been formatted more appropriately. The following changes have been made:

- the font size has been increased;
- the header row is made clear by setting the text in bold and lightly shading the row;
- the internal vertical lines have been deleted to allow patterns between the rows to be more easily seen;

13.2.5 Design the layout

The fifth step is to design the layout. The layout needs to be designed so that a person viewing your poster can quickly understand the sequence of the information. This is

FIGURE 13.16 A more thoughtfully formatted table.

Arsenic species	Formula	Concentration/ppm
As(III)	$As(OH)_3$	5
As(V)	$AsO(OH)_3$	11
MMA(V)	$CH_3AsO(OH)_2$	0
DMA(V)	$(CH_3)_2AsOH$	0
AsB	$(CH_3)_3As^+Ch_2CH_2COO^-$	0

most easily achieved using appropriate and conventionally ordered headings, as shown in Figure 13.17.

In addition to using a conventional heading sequence it is also helpful to use a design grid, as shown in Figure 13.18.

A design grid is simply a series of horizontal and vertical lines on the page that are used for lining sections up; the lines are then removed when the poster is printed. Most drawing software has a design grid feature, or alternatively you can draw the lines manually with a line tool to create the grid and remove them before printing.

In the above examples the methods section is unusually long. However, remember this is a poster summarizing a current research technique. More commonly, the methods section would be much shorter, leaving space for fuller results, discussion, and conclusion sections. In spite of the unusually long methods section, however, the above layouts are fairly conventional; following a symmetrical three-column format. There are many other alternatives though, and you will need to choose the most appropriate one to fit your content. Other examples for both landscape and portrait layouts are shown in Figure 13.19 and Figure 13.20.

FIGURE 13.17 Use of appropriate headings to show sequence.

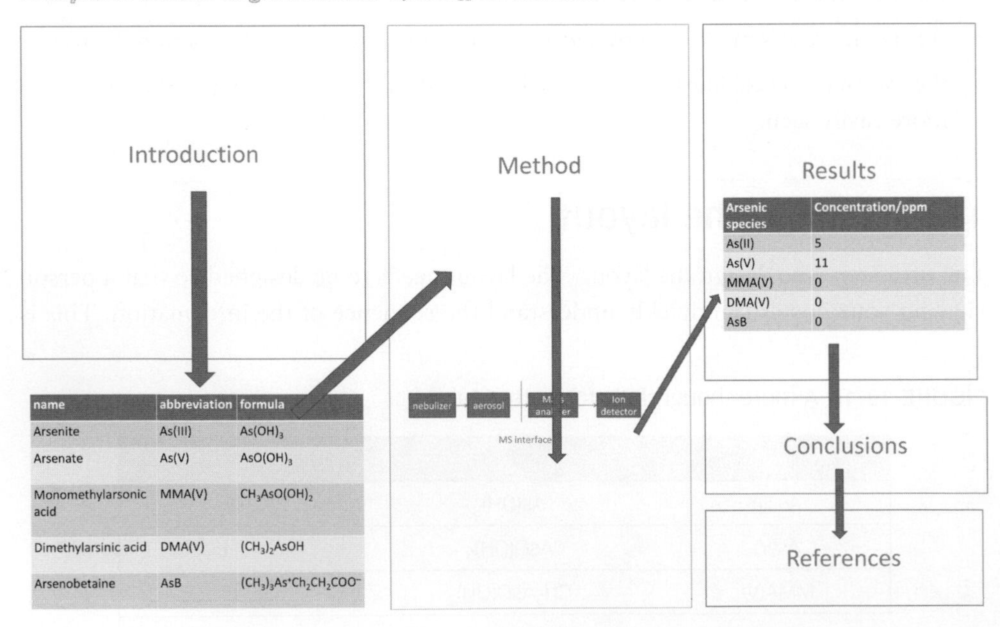

FIGURE 13.13 Use of a design grid to show sequence.

Analysis of arsenic in groundwater by Joe Bloggs and Teresa Green

Introduction

Arsenic is a metalloid that can be found in the environment in rocks, soil, and water. It occurs naturally as the minerals arsenopyrite FeAsS), realgar (AsS), and orpiment, (As2S3). Levels of arsenic in the soil can be found to be from 0.1 to 40 ppm. Arsenic can enter groundwater by erosion, dissolution, and weathering. Other sources include geothermal springs and volcanic activity. It can also be released by brines that are used to produce natural oil and gas, as a by-product of gold and lead mining and the combustion of coal. Industrial contaminations occurs through production of lead-acid batteries, paper production, and glass and cement manufacturing.

Arsenic occurs in drinking-water as As(III) or As (V).). Organic arsenic species are concentrated in seafood but are much less harmful to health than inorganic arsenic compounds as they are readily eliminated by the body. The European Union and the World Health Organization recommend arsenic limits in drinking water of 0.01 ppm. Arsenic poisoning or arsenicosis results in many ailments such as skin cancer, cancers of the bladder, kidney and lung, gangrene, diabetes, high blood pressure, and reproductive disorders. It is obviously important to be able to detect low levels of arsenic in aqueous sample.

Table 1 Common arsenic species in groundwater

name	abbreviation	formula
Arsenite	As(III)	$As(OH)_3$
Arsenate	As(V)	$AsO(OH)_3$
Monomethylarsonic acid	MMA(V)	$CH_3AsO(OH)_2$
Dimethylarsinic acid	DMA(V)	$(CH_3)_2AsOH$
Arsenobetaine	AsB	$(CH_3)_3As^{-}CH_2CH_2COO^{-}$

Method

Groundwater contains a variety of inorganic and organic arsenic species and these must be separated by chromatography before determination. Arsenic can be determined by a variety of analytical methods including colorimetry, atomic absorption spectrometry, XRF, ICP, and ICPMS. In this project the arsenic species were separated by ion exchange chromatography and measured by ICPMS. A Dionex ion-exchange column was used with gradient elution with 2 mM tetramethylammonium hydroxide and 10 mM ammonium carbonate binary mobile phase. Six arsenic species were separated over a period of 20 min in the order AsB, DMA(V), MMA(V), As(III) and As(V).

A Shimadzu inductively coupled plasma mass spectrometer with a miniaturized torch was used for the detection. The eluent from the ion exchange column was introduced into the plasma by a nebulizer.

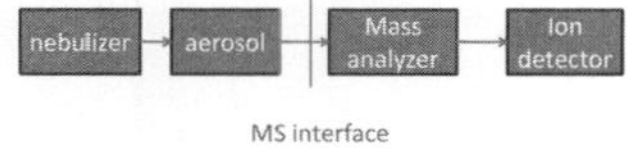

Figure 1 Schematic representation of ICPMS

Standards of each of the arsenic compounds were prepared in water over the range0-50 ppm. Six-point calibration curves for the arsenic species were obtained by plotting the peak areas against the concentration of arsenic for each species

Results

The calibration curves gave corellation coefficients in excess of 0.999. The detection limits ranged from 0.071 to 0.40 ppm.

The concentration of each of the arsenic species in the groundwater sample are given in table 2.

Table 2 Results

Arsenic species	Concentration/ppm
As(II)	5
As(V)	11
MMA(V)	0
DMA(V)	0
AsB	0

Conclusions

As effective method for determination of arsenic in groundwater was developed,. The levels of organoarsenic compounds was below the recommended limit. However, there was significant inorganic arsenic present.

References

B.K. Mandal, O. Yasumitsu and K.T. Suzuki, Identification of dimethylarsinous and monomethylarsonous acids in human urine of the arsenic-affected areas in West Bengal, India, *Chem. Res. Toxicol.* , 2001, **14**, 371-378.
Z. Gong, X. Lu, M. Ma, C. Watt and X. Le, Arsenic speciation analysis, *Talanta*, 2002, ***58***, 1, 77-96
T. Nakazato, T. Taniguchi , H. Tao , M. Tominaga and A. Miyazaki, Ion-exclusion chromatography combined with ICP-MS and hydride generation-ICP-MS for the determination of arsenic species in biological matrices, J. Anal. At. Spectrom., 2000, 15, 1546-1552

FIGURE 13.19 Alternative landscape layout.

FIGURE 13.20 Alternative portrait layouts.

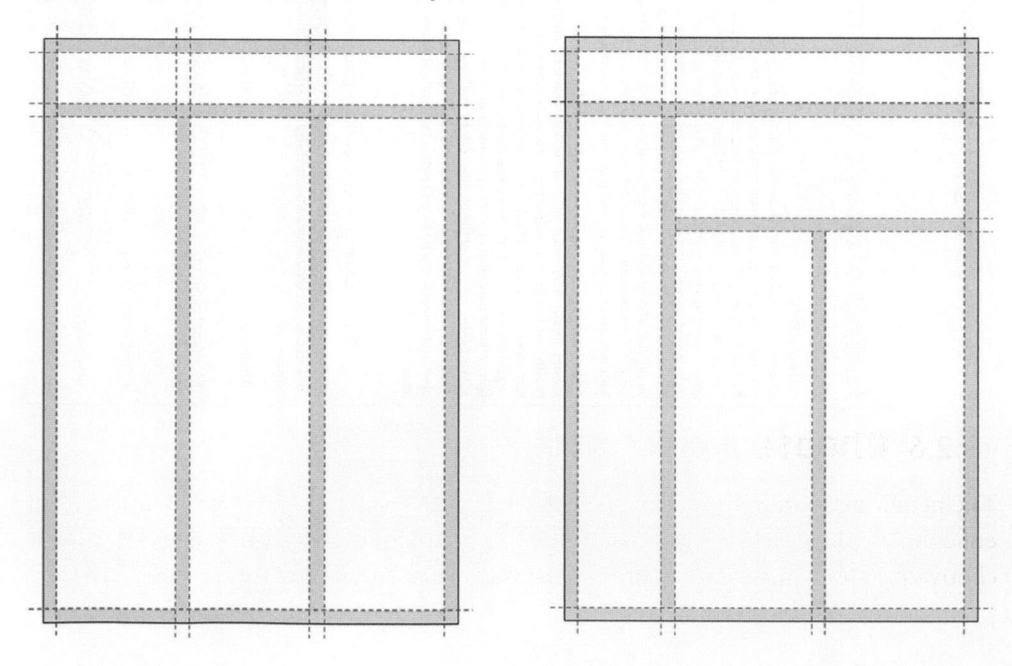

BOX 13.2 Do the small details really matter?

Posters can be very fiddly and therefore time consuming to create. However, remember the academic context of poster presentations that we summarized in Section 13.1: in a busy conference setting with lots of posters to view, people make very quick (often unfair) judgements about whether or not they will bother viewing a poster. We said that a poster needs to stand out from the crowd in some way if it is going to be viewed. Positively, this will include a snappy title, good visual impact, and information that is readily understandable; negatively small errors in the design can put people off and lead them to think that if the design isn't accurate the data might not be either. Also, if you are creating a poster at A4 size then enlarging it to A1 or A0, any small errors at A4 will be enlarged 300 per cent or 400 per cent respectively. Therefore attention to detail is important. See Figure 13.21 for an example of how small inaccuracies with the layout at the A4 stage lead to much larger inaccuracies at A0.

FIGURE 13.21 How small errors are noticeable when the design is enlarged to poster size.

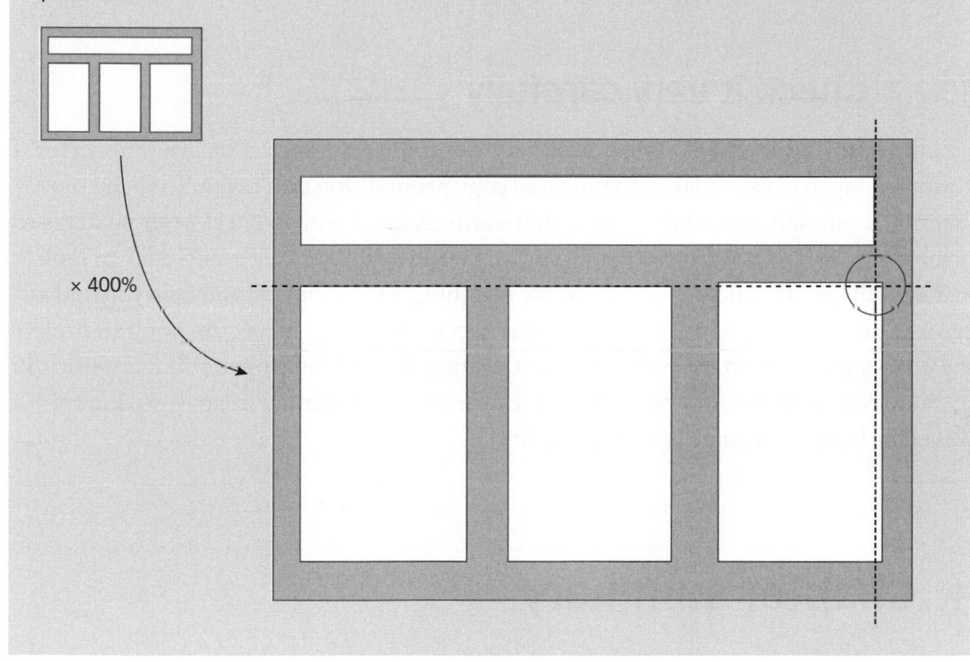

13.2.6 **Choose a colour scheme**

The penultimate step is to choose a colour scheme. Try to use only two to three different colours, plus black, which is always best for the smallest text. There are two ways of choosing your colours: you can use colours from your images or a colour wheel.

Colours from images

One way of choosing colours is to pick a colour that is represented in your images. So, for the poster we have used as an example we might choose red and grey because of the molecular structure.

Colour wheel

The alternative to choosing colours from images (either because you don't have any appropriately coloured images or you are not confident in choosing from them) is to choose colours from a colour wheel. A colour wheel, as shown in Plate 1 shows the primary, secondary, and tertiary colours and can be used to choose appropriate mixes of colour.

You can choose:

- analogous colours (colours next to each other on the colour wheel);
- complementary colours (colours opposite each other on the colour wheel);
- or shades (different shades of the same colour).

13.2.7 **Check it very carefully**

Finally, you need to check your poster very carefully. This is particularly important if you are going to get an enlarged, laminated copy printed: you don't want to spend money printing it out only to find the errors afterwards. A good way to check your poster is to print it out A4 (preferably in colour), stick it on the wall, and take a step back to look at it. This will simulate more closely how the real thing will be viewed and help you find any errors more easily. Alternatively, if you have access to a data projector, you can project the image onto a screen to see the effect of the image being blown up to full size (although the image resolution is much less for digital images than printed images, so don't use a projected image as a guide for image quality).

* Chapter summary

Designing academic posters can be a difficult and time-consuming task. However, by understanding the purpose of posters and the key steps involved in creating them, designing and producing quality academic posters becomes an achievable goal. Designing posters is a particularly useful skill if you continue studies beyond undergraduate level as they are a common form of presenting research at academic conferences. When producing your poster remember the key steps:

- plan the academic content carefully;
- cut the text down to between 300 and 500 words;

- choose an appropriate font;
- prepare your tables, structures, schemes, diagrams, and images;
- design the layout;
- choose a colour scheme;
- check it very carefully.

Chapter 14

Getting the most out of revision

➲ Introduction

The results you gain for your degree will depend on how you perform in the various forms of assessment that you are required to undertake. We have already addressed essays, practical reports, and presentations; the one we haven't dealt with yet directly (and the one that for most degrees accounts for the majority of the marks) is examinations. In order to do well in exams you have to do well at revision (two words that usually conjure up fairly negative emotions in people). This chapter will show you how you can revise well; Chapter 15 will deal with the exam skills.

If exams account for the majority of the marks that make up your degree result, it is understandable that people feel pressured or stressed by them. One important way of significantly increasing your chances of doing well and of alleviating some of the pressure and stress of exams is to get the revision part right. If you can go into an exam confident that you have prepared yourself as well as you could have done (or, at least, reasonably well), then a significant amount of the pressure and stress can be eased, making it more likely that you can perform well. However, getting the revision part right takes a significant amount of organizing; it is no good leaving it to the last minute.

First though, let us consider what revision actually is. In very general terms revision is to do with going over a subject again, usually in preparation for some kind of exam. But exactly how you go over a subject again is critical to whether or not the revision is effective. You will know if your revision is effective because revision that is effective will make the material you are revising easier to:

- understand;
- remember;
- apply.

This chapter will give you some tips and techniques to help you make this happen.

14.1 Get yourself organized!

We have already highlighted the fact that getting revision right takes a significant amount of organizing. However, getting yourself organized is easy to say but is more difficult to actually do, especially when you're beginning to feel the pressure of exams looming and you just want to get on with revising. It is important, however, to see the organizing or planning aspects of revision as part of the process, rather than something to get out of the way so you can get on with the 'real work'. This is similar to how we identified planning writing as an important part of writing essays in Chapter 7. Time spent *organizing* revision is time well spent (as long as you don't use it as an excuse to procrastinate!).

There are a number of questions you need to ask yourself in order to get organized.

- What will I be examined on?
- Do I have all the necessary material?
- How much time do I have before the exams?
- How am I going to allocate my time?

These questions will form the structure of this first section.

14.1.1 Find out what you will be examined on

First, then, you need to find out as much as possible about what you will be examined on. This includes both the format and scope of the exam (or exams). In terms of format this would include finding out answers to the following questions.

- How long will the exam last?
- How many questions will there be?
- Will you have any choice about which questions you answer?
- What type of questions will you be asked (for example multiple choice, short answer, or essay)?

Knowing the answers to these basic questions will help you to know what to expect. The better informed you are the better you can prepare. It also means that you are not surprised when you open the exam paper and so will be more able to concentrate on answering the questions rather than spending time trying to understand issues of format.

In addition to the format of the exam you also need to consider its scope. What are you expected to know in terms of both breadth (the range of information) and depth (the amount of detail)? Imagine you were revising for an exam on *Advanced analytical techniques*, if you had revised enough in terms of breadth but had revised in insufficient depth your knowledge of analytical techniques would be too *shallow*. For example, you may have revised atomic and molecular spectroscopy, chromatography, and electrochemical methods, but failed to learn them in sufficient detail. Alternatively, if you had revised enough in terms of depth but had revised insufficient breadth your knowledge of analytical techniques would be too *narrow*. For example, you may have revised chromatography in intricate detail, but if you failed to learn about other techniques as well (atomic and molecular spectroscopy and electrochemical methods), then you would be very limited in the number of questions you can answer.

This may all seem fairly obvious, but finding out what you will be examined on is an important and often overlooked area of revision. Sometimes it is overlooked due to carelessness, other times due to a deliberate but irrational avoidance mentality. Whatever the reason, it is not difficult to fix. There are a number of ways you can find out information about exam format and scope including:

- module handbooks;
- past papers;
- course tutors.

Each of these will give you important insights into what you can expect.

14.1.2 Check you have all the necessary material

The second question to ask in order to get yourself organized is 'Do I have all the necessary material?' When you have found out about what you will be examined on (the breadth and depth) you then need to look at the material you have and decide whether or not you have enough. It is important that you think about more than just your lecture notes at this point. Other sources of important information include:

- notes from tutorials;
- reports or essays you have written for coursework;
- lab reports that you have submitted;
- notes from presentations that you have been required to give;
- notes from additional reading for all of the above;
- feedback from tutors on all of the above;
- a list of the module learning outcomes.

As you can see, the amount of information you will have accumulated over a term or semester will be considerable. At this stage you are not supplementing your notes, simply auditing them to check where there are any gaps. We said earlier that module handbooks were an important source of information on exam format and scope, and as such they help you to identify gaps in your notes and so help you determine whether or not you have enough material. Simply make a list of what information you have on a summary sheet and compare it to what you are expected to know, as described in the learning outcomes for the particular module you are revising for. Learning outcomes describe what you should be able to do at the end of the module and you should be able to demonstrate them during assessment, for example in an examination.

When you identify gaps in your material (we say 'when' rather than 'if' because it is unlikely that you will have everything) make a note of what the gaps are so you can come back to them later. It is also worth identifying whether the gap is a breadth issue (an insufficient range of information) or a depth issue (insufficient detail). The other factor to consider is whether certain gaps actually matter: if the format of the exam allows you some choice as to which questions you answer (for example, some essay-based exams) you may not need to revise everything. It is usually better to have revised most things in sufficient depth rather than everything at just a surface level. It may also be that there were some sections of the module or course that you just could not get to grips with and still don't understand. There are risks associated with this strategy of selective revision, so you need to be careful. For essay-based exams where you have a choice of questions, it is still very important that you revise more topics than you actually have to answer questions on, so that you do have a choice when you come to read the paper. A reasonable rule of thumb is always to learn at least two more topics than you have to answer questions on, so if, for example, your typical essay paper requires you to answer three questions, then it is best to know five topics really well.

14.1.3 Note how much time you have before the exams

The third aspect of getting yourself organized is to be clear about how much time you have to revise. Once you know the dates of your exams you will know how much time is available to you. The best way to impress on yourself how much time you have available is to represent the time visually by using a diary, calendar, or simply a sheet of paper. In addition to the exam dates themselves, you need to include other fixed points, such as any remaining coursework deadlines, the final lecture, and any non-course commitments you have, such as holidays. All of this needs to be represented on a planner, such as the one shown in Figure 14.1.

It is unlikely that all of your exams will be at the same time, so you can stagger the start of the revision for each exam accordingly. Knowing when your exams are will help you to appreciate how much time you have before the exams. The next step is to decide how you are going to allocate the time you have.

14.1.4 Decide how you are going to allocate your time

Deciding how to allocate your time is an important element of getting yourself organized. The following guidelines will help.

Decide what you are going to revise and how much

In order to decide how you are going to allocate your time you need to decide what you are going to revise and how much you are going to revise it. You will already have some idea of this having found out what you will be examined on (see Section 14.1.1) and whether

FIGURE 14.1 Planning your time.

April						
		1	2	3	4	5
6	7	8	9	10	11	12
13	14	15	16 Final lecture	17	18	19
20	21	22 Essay due in	23	24	25 Away for	26 weekend
27	28	29	30			

May						
				1	2	3
4	5	6	7	8	9	10
11	12	13	14 Exam 1	15	16 Exam 2	17
18	19	20 Exam 3	21 Exam 4	22	23	24
25	26	27	28	29	30	31

you have all the necessary material (see Section 14.1.2), but now you need to make some definite decisions about how you will allocate the revision of your material to the time you have available. There are some important considerations to bear in mind at this point:

- if you are going to know your material in sufficient depth you will need to revise subjects more than once;
- however, that doesn't mean you have to revise absolutely everything (remember, it is usually better to have revised most things in sufficient depth rather than everything at just a surface level);
- you might also decide that you need to revise more for some exams more than others – you don't necessarily have to allocate equal amounts of revision time for each exam.

A useful process is to make an overview of each subject, listing the topics in that subject, and the headings under each topic (you will already have most this information if you have identified whether you have the necessary material). This creates an index of what needs to be revised so that you can divide the revision into easy-to-manage sections. Once you have this overview, you are ready to put together your revision timetable.

Make a revision timetable

The kind of plan represented in Figure 14.1 gives you a useful overview but it doesn't give you enough detail about how you need to allocate your time. You need to zoom in on this overview to create monthly, weekly, and even daily plans of what you need to do.

Imagine your four exams were as follows:

1. Introduction to thermodynamics and kinetics – 14 May;
2. Main group chemistry – 16 May;
3. Functional group chemistry – 20 May;
4. The analytical approach – 21 May.

According to your plan (Figure 14.1) this gives you four weeks between your final lecture (16 April) and your first exam (14 May). However, there is also your final essay to submit on the 22 April, and assuming you will need the time between the final lecture and the submission date to work on the essay, that leaves just over three weeks to revise before your first exam. The remaining three exams then fall within the following week. These weeks need planning carefully if you are to cover the necessary material in the time available; this is where the weekly and daily plans come in (Figure 14.2).

> **Try this: Planning your revision**
>
> Two months before your next set of examinations prepare a detailed revision timetable. Make it realistic by taking into account other commitments that you might have, such as work, social events, and family time.

FIGURE 14.2 Monthly, weekly, and daily revision plans.

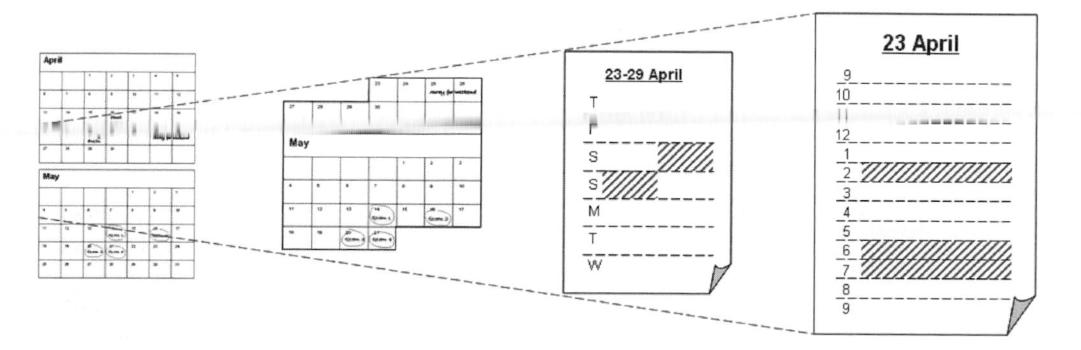

Be realistic and build in some slack

It is important to be realistic in your planning. One of the reasons for getting a clear view of how much time you have is to help you not to panic and so end up trying to cram too much revision into a short space of time. If you are clear how much time you have and what you need to do, you can then allocate your time realistically. A realistically planned revision timetable will have some slack built in so that when your plans need to change (for instance due to a certain topic taking longer than expected) there is space for this to happen. Of course, if the slack time is not needed you can use it for other things, for instance going over a topic again in more detail or perhaps rewarding yourself with an unexpected break. It is difficult to be precise about how much slack time is needed, but if you planned to revise seven hours in a day, you could leave the seventh hour unallocated to use for whatever was necessary. An example is shown in Figure 14.3.

In addition to planning in slack time you also need to plan in breaks: revising non-stop for seven hours is not an efficient use of time. The longer you go without a break, the more difficult it becomes to concentrate and absorb information. Breaks from revision, therefore, are not a luxury; they are a necessity if you are to be able to revise in a sustainable manner. Therefore, the morning and afternoon slots represented in Figure 14.3 shouldn't be three hours without a break, but rather shorter blocks with breaks built in. Keep your breaks short and free from distractions; just getting up to walk round

FIGURE 14.3 Daily plan with slack built in.

27 April

9..................................

10...main group chem.........

11...main group chem..........

12...main group chem..........

1.....lunch.........................

2.....meet Holly.................

3....thermodynamics...........

4....thermodynamics...........

5....thermodynamics...........

6....thermodynamics...........

7...tea..............................

8...slack...........................

9....................................

the room or make a drink is enough to renew your concentration. You also need to decide what times of day you work best; some people find working in the mornings easier, others find evenings more productive. Whatever your preference is, try to use it to your advantage, but don't use it as an excuse not to get up in the morning!

Try this: Planning your revision in detail

Using your overall revision plan as a guide, produce a detailed and realistic daily revision plan that you are comfortable you could stick too.

Do other stuff

Obviously it is important to be focused on your revision, but it is important to have a life outside of revision too. The longer your revision lasts the more important it is to plan in time to do other things. Planning in time for other things isn't simply an end in itself but can be used to help you in your revision in a number of ways, for instance:

- provide you with a break so you can come back to your revision more mentally refreshed;
- re-energize you so you can come back to your revision more physically refreshed;
- act as a reward for having achieved a certain target in your revision.

Precisely what you do to achieve these benefits will depend on what you find helpful. People often find exercise helpful, or perhaps spending time socializing. You will need to be disciplined and not get carried away, but a short period of activity, perhaps each day, that you find enjoyable and would achieve some of the benefits listed above will help keep your revision on track and sustain your motivation.

Use variety

In addition to spending short periods of time doing other things, another helpful strategy is simply to use a variety of approaches to your revision. You will know that this is helpful if you have ever tried to spend a revision session just reading, for example. Spending a large amount of time only using one approach can quickly become tedious and demotivating. But this isn't necessarily a sign that you need a break, it might just be a sign that you need to change your method. Try testing yourself with some questions or flash cards, or drawing a diagram, or writing a summary of what you have just read. We will deal with these, and other, suggestions in more detail shortly when we consider active revision techniques, but the important point to note here is the principle that using a variety of approaches to how you revise can be helpful. A change is often as good as a break (and better than a break if you are using breaks to avoid getting down to revision!).

Go public

Lastly in this section on allocating your time, consider making your revision timetable public; you could stick it up on a kitchen cupboard or on your bedroom door. There are a number of reasons this can be helpful:

- it reminds you of what your plans are;
- it informs others who you share your accommodation with what your plans are;
- it makes clear to others when you are available and when you are not;
- it makes you accountable to others ('I thought you were supposed to be revising?' or 'Shouldn't you be taking a break now?').

Whether or not you want to make you plan public will depend on who you live with, but it is definitely worth considering.

14.1.5 Check your environment is suitable

Finally, in this section on getting yourself organized, you need to think about the environment in which you will revise. You might be the sort of person who always prefers to work in the same place every time, or perhaps you like variety. Some people find the campus library a helpful place to revise, whist others prefer to study in their own room. Whatever your preferences try and make decisions based on what will help your revision to be most effective. A suitable studying environment is one that is free from distractions. For example, it should be:

- well lit;
- quiet;
- not too hot or too cold;
- reasonably tidy and free from irrelevant clutter.

Sort these things out quickly before your revision session begins, but don't use them as an excuse to avoid starting!

14.2 Use active revision techniques

We said at the beginning of this chapter that you will know when your revision is being effective because it will make your material easier to understand, remember and apply; this is what active forms of revision help you to do. **Active** revision is any form of revision that makes you interact with the material in an involved and thoughtful way; this includes condensing your notes, drawing summary diagrams, and testing yourself with questions or flash cards. **Passive** forms of revision, however, such as reading or copying, lack the focus provided by active forms and are more difficult to sustain for longer periods.

So, now you're organized, what are you actually going to do when you sit down for your first revision session? Assuming that you have planned your revision timetable so you have a reasonably clear idea of the topics you will be covering on any given day and how long your revision session for the day concerned will last, you are now in a position to begin to work on the detail. This will involve:

- filling in any gaps in your material;
- condensing your notes carefully;
- reviewing your notes regularly.

14.2.1 Fill in any gaps

The starting point then is to fill in any gaps in your material you have identified (as described in Section 14.1.2) and have decided need addressing. Your material could be supplemented from a number of sources: if you have simply missed a lecture, then borrowing a friend's lecture notes may be adequate, but it is more likely that you will need to do some additional reading. This needs to be done in a certain sequence: make sure you understand the basics first before you progress onto more advanced issues. In practice this means you need to start with lecture notes and handouts, followed by key chapters from core texts on your reading list. If you still need more detail then, once you are confident you understand the basics, you could move onto more specialist publications. This is where it is important that you have understood what you will be examined on (see Section 14.1.1) so you are aware how much additional reading is necessary. There will always be more reading that you could do, the question you need to answer is whether there is more reading that you must do? At this stage, you need to be strategic in your approach: you don't have a lot of time to be undertaking detailed additional reading, therefore you need to identify exactly what is necessary and do that, but avoid doing more.

14.2.2 Condense your notes carefully

Once you have filled in any gaps in your notes you will probably have a lot of material. Therefore, a vital stage in any revision strategy is to condense your notes into a format that is more manageable. There are two simple reasons for this:

- the process of condensing your notes helps you to learn the material;
- the end product of condensing your notes provides you with a summary of your material that you can easily review (also enabling you to learn your material).

We will address the review part shortly (Section 14.2.3), but first we will deal with the condensing element. Condensing your notes is an example of an active form of revision: it forces you to make decisions about which parts of your material are the most important and will help you to check how much you understand. This process consolidates your knowledge and helps you to make connections between different aspects of your learning and identify underlining principles. But how do you actually condense your notes? Condensing notes involves the following stages (and is illustrated in Figure 14.4).

- Take your original notes on a topic (from various sources) and make a condensed version, perhaps on several sheets of paper.
- Then take this condensed version of your notes and try to condense the information still further, perhaps to a single sheet of paper.
- Finally, write an overall summary of the particular topic you are revising, perhaps on an index card.

The principle of condensing is simple enough, but how do you decide what information is sufficiently important for you to record and what you can leave out? Clearly you don't want to simply copy everything out (this would be a passive form of revision that would quickly become ineffective) but, equally, you don't want to be so sketchy with your notes that when you come back to them at a later stage there is insufficient detail for you

FIGURE 14.4 Condensing notes.

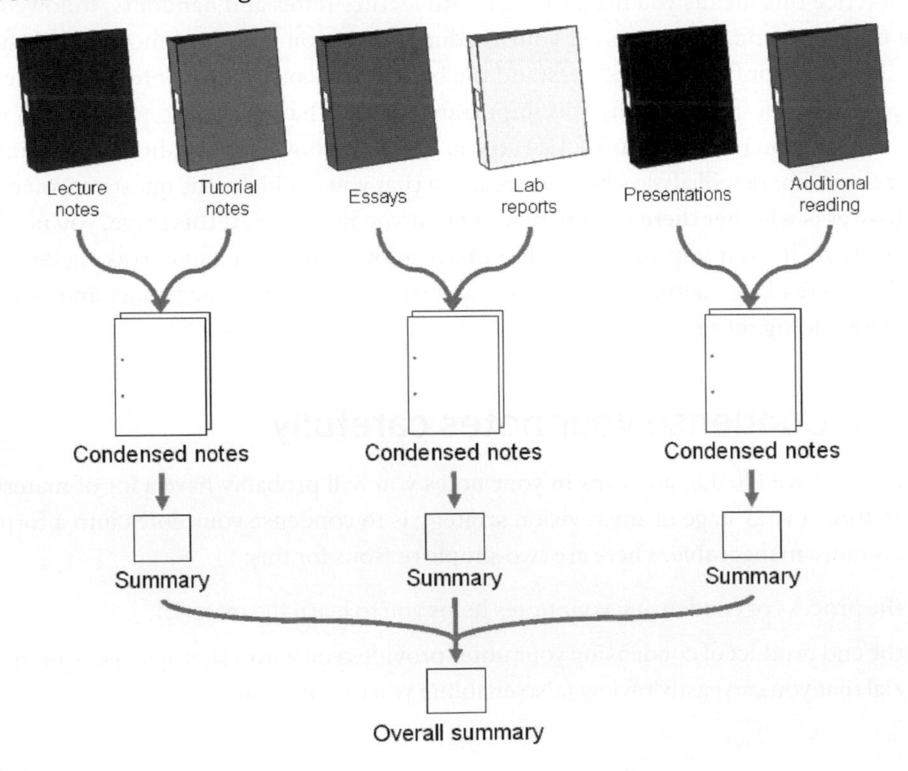

to work from. This is why this process involves a number of stages: you can't skip from your lecture notes straight to an overall summary without going through the intervening stages. The reason for this is that, in producing condensed notes and then summaries of these condensed notes, you are not only producing a more condensed version of your material on paper but also structuring and connecting the information in your mind and making decisions about what is sufficiently important to record and what you can leave out. At each stage you are attempting to distil out the essence of the particular topic you are studying in order to understand it better. What you will be left with, therefore, is not everything you need to know on the topic but sufficient information to prompt you to recall the source information from which it was distilled.

Exactly how much information you record depends on how much you already know and so will vary from individual to individual. What you are aiming for, however, is to pick out key facts and create structures to help you to understand whole concepts. The following example illustrates how this might work.

Imagine you had the following notes from a lecture on molecular orbital theory.

We assume that that we can construct a reasonable first approximation to the molecular orbital by superimposing atomic orbitals contributed by each atom. This modelling of a molecular orbital in terms of contributing atomic orbitals is called the *linear combination of atomic orbitals* (LCAO) approximation. A 'linear combination' is a sum of wavefunctions with various weighting coefficients. In simple terms, we combine the atomic orbitals of contributing atoms to give molecular orbitals that extend over the entire molecule.

In the most elementary form of MO theory, only the valence-shell atomic orbitals are used to form molecular orbitals. Thus, , the molecular orbitals of H_2 are approximated by using two hydrogen 1*s* orbitals, one from each atom, *A* and *B*:

$$\psi = C_A\varphi_A + C_B\phi_B$$

In this case the **basis set**, the atomic orbitals ϕ from which the molecular orbital is built, consists of two H1*s* orbitals, one on atom A and the other on atom B.
The probability of finding an electron in the volume of space is ψ^2

$$\psi^2 = C_A{}^2\varphi_A{}^2 + 2C_AC_B\phi_A\phi_B + C_B{}^2\phi_B{}^2$$

The coefficients *c* in the linear combination show the extent to which each atomic orbital contributes to the molecular orbital: the greater the value of c^2, the greater the contribution of that orbital to the molecular orbital.

Important points to be noted:

1. *N* molecular orbitals may be constructed from *N* atomic orbitals.
2. The same rules that are used for filling atomic orbitals with electrons apply to filling molecular orbitals with electrons, i.e. the build-up principle, Hund's rule, Pauli exclusion principle.

FIGURE 14.5 Annotated notes.

We assume that that we can construct a reasonable first approximation to the molecular orbital by superimposing atomic orbitals contributed by each atom. This modelling of a molecular orbital in terms of contributing atomic orbitals is called the linear combination of atomic orbitals (LCAO) approximation. A 'linear combination' is a sum of wavefunctions with various weighting coefficients. In simple terms, we combine the atomic orbitals of contributing atoms to give molecular orbitals that extend over the entire molecule.

In the most elementary form of MO theory, only the valence-shell atomic orbitals are used to form molecular orbitals. Thus, the molecular orbitals of H_2 are approximated by using two hydrogen 1s orbitals, one from each atom, A and B:

In this case the basis set, the atomic orbitals ϕ from which the molecular orbital is built, consists of two H1s orbitals, one on atom A and the other on atom B.

The probability of finding an electron in the volume of space is ψ^2

The coefficients c in the linear combination show the extent to which each atomic orbital contributes to the molecular orbital: the greater the value of c^2, the greater the contribution of that orbital to the molecular orbital.

Important points to be noted:

1. *N* molecular orbitals may be constructed from *N* atomic orbitals.
2. The same rules that are used for filling atomic orbitals with electrons apply to filling molecular orbitals with electrons i.e. the build-up principle, Hund's rule, Pauli exclusion principle

A useful way to start to summarize such notes is to annotate them as you read them, as shown in Figure 14.5.

You could also choose to summarize the content in your own words, as shown in Figure 14.6.

FIGURE 14.6 Summary notes.

Superimpose atomic orbital to form molecular orbitals

Linear combination of atomic orbitals LCAO

Sum wavefunctions with different weightings

Get MOS over all molecule

For hydrogen 1s orbitals, one from each atom, A and B:

Basis set

C mixing coefficient

N molecular orbitals gives *N* atomic orbitals.

FIGURE 14.7 Summary diagram.

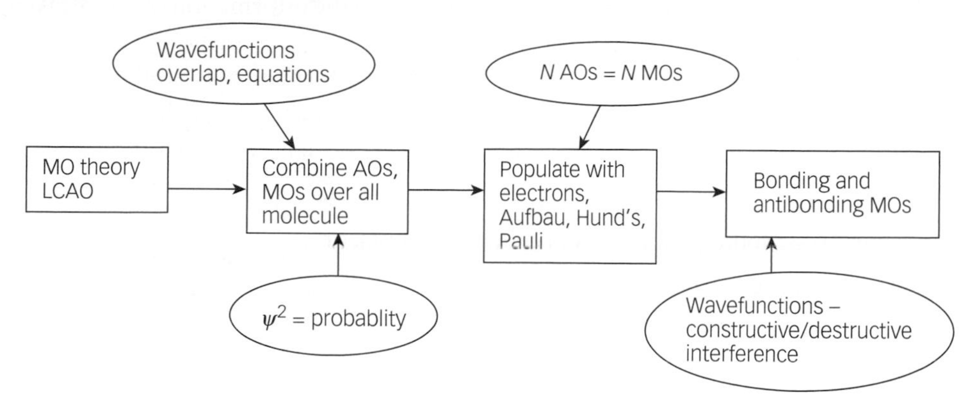

To condense the information further you could create a summary diagram, as shown in Figure 14.7. This summary diagram has the added advantage of including additional information too.

In addition to considering the content of the information you are condensing you also need to consider its format. The example used in Figure 14.7 is a deliberately visual example. This is because many people find it easier to remember something that is visual rather than something that is essentially text based. In Chapter 2, *Making the most of lectures*, we suggested a number of points to improve your note making, many of these are also relevant for note making for revision. Remember to try to use:

- headings and subheadings;
- colour;
- space;
- figures, including labelled diagrams and flow charts.

14.2.3 Review your notes regularly

We observed above that condensing your notes provides you with a summary that you can easily review. If you have ever tried to review uncondensed notes you will know that it is a very difficult exercise to sustain because of the sheer volume of information you are trying to read. Even if you manage to review uncondensed notes for a particular topic, the more topics you have the more unsustainable it becomes. If you have produced summaries, however, not only do you have your notes in a handy format for you to review, as identified above, the process of condensing your material will have helped you to learn it. Clearly you won't remember everything (that is why the review stage is important) but you will remember some and you will also have something that is *reviewable*.

The importance of reviewing your notes regularly can be highlighted by considering what happens when you don't review your notes regularly. What inevitably happens is that the longer you go without reviewing your material the more you forget. Without regular reviews of what has been revised you won't retain the information you have taken the trouble to learn.

The relationship between reviewing and recalling information is straightforward enough, but *how* do you actually review your material? There are many different methods of reviewing information but whichever method you choose it needs to test your recall of a topic by:

- checking your ability to understand central concepts; or
- remembering key facts; or
- linking the information to other topic areas;
- or perhaps all three!

As we have already noted, you won't remember absolutely everything, and you can always go back to your original notes to remind yourself of particular details, the important thing is to do the reviewing. Methods of review include:

- testing your recall of key facts with index cards (or flash cards), putting to one side the cards you can remember and repeating the cards that you can't;
- reproducing from memory headings and keywords for a learnt topic area and then checking them against your notes;
- recording yourself explaining concepts or theories to an imagined audience, then play back your explanation, checking it against your notes; making a visual summary (like the flow chart in Figure 14.7 to summarize information you have remembered;
- reproducing important diagrams that summarize large amounts of information or whole concepts, for example metabolic pathways;
- enlisting the help of a friend to test your recall using any of the above methods;
- putting up summaries of key information around your home to give you the opportunity for on-the-spot tests;
- testing yourself with question banks and past papers (more on this shortly).

Reviewing your material need not take a lot of time. A good time to review material is at the beginning of a revision session but it can also be fitted into any spare time during the day. Frequently reviewing your material will help to build up your confidence and reassure you that you are making progress.

14.3 **See the big picture**

We have noted above that condensing your notes supports the consolidation of your knowledge and helps you to make connections between different aspects of your learning and identify underlining principles. However, when paying attention to detail, it is easy to lose sight of the context and principles. It is important, therefore, that you make a special effort to appreciate the big picture. Try not to get so focused on the details that you fail to see either where the information fits in to the larger context or perhaps a simple principle that underpins it, both of which would make it easier to understand and remember.

The exact parameters of the big picture will depend on how much breadth and depth you are required to know, and this will change depending on the module choices you make throughout your course and the level of study you are at. However, context and principles are important to appreciate and identify because they help you to develop a means of structuring, storing, and retrieving information. They also help because exam questions rarely ask the precise question you had been hoping for; rather they require you to apply your knowledge to slightly different situations.

Imagine walking into a room with papers scattered across the floor and then being asked to find a particular piece of information – it would be very difficult! If you did manage to eventually find the correct piece of paper it would be difficult to then explain how the information on the piece of paper relates to all the other information on the pieces of paper still on the floor. However, imagine the same papers in the same room but this time organized appropriately into a filing cabinet. Suddenly it becomes much easier to find the required information and relate it to other information. Seeing the big picture is a bit like storing papers appropriately in a filing cabinet (or information in a database): it enables you to appreciate how information fits into a wider context and identify the principles that underpin it. Without these elements the information you are trying to learn will remain abstract and unrelated, which makes it much more difficult to understand, remember, and apply. It is important therefore that you spend some time standing back from the detail to think about structure. Appreciating context and identifying principles are two specific ways of doing this, but anything that helps you to reflect on and question the information you are revising, helping you to understand themes and relationships rather than detail, will help you to see the big picture.

14.4 Practise outputting the information you have learnt

We have noted the importance of understanding, remembering and applying the information you learn. However, the steps we have focused on so far are largely concerned with the first two elements: understanding and remembering. If you are going to perform well in the exam room (more on this in the next chapter) then you need to be able to not only understand and remember information but, importantly, you also need to be able to apply it. This is because exam questions, whatever the format, are specific and you are therefore required to give specific answers. When you open an exam paper you are very unlikely to find a question which reads 'Please tell us everything you know about the chemistry of Group 13 and Group 17 elements'. It is much more likely that the question will be more like 'Compare and contrast the chemistry of the Group 13 and Group 17 elements'. Therefore, the information you have taken time to understand and remember needs to be applied to the specific question. It would be very unfortunate if you had taken the time and effort to understand and remember information but failed at this last crucial step. The best way to improve your ability to answer questions specifically is to practise. The two main ways of practising are:

- creating question banks; and.
- using past papers.

14.4.1 Create question banks

A question bank is simply a collection of questions that you create that you think you might be examined on. The questions shouldn't be full essay questions, but short, one-line questions that test very specific knowledge. Question banks are useful for a number of reasons:

- past papers are not always available to you;
- the format of the exam might have changed;
- thinking up questions helps you to reflect on how much you know;
- answering the questions helps you to apply your knowledge.

Question banks can be created from any questions that you think you might need to know the answer to. A good place to start is the learning outcomes for the modules: use these as a prompt to help you think of questions. Keep adding questions to your question bank throughout your course; questions prompted from lectures, or tutorials, or reading. You can then use these questions to check your knowledge of the material and so monitor your progress, or as a means of review, or to supplement your use of past papers. It is also beneficial to team up with some fellow students to share the task of creating a question bank; this will save you time as individuals and give you a greater variety of questions than you might have created on your own.

14.4.2 Use past papers

Past papers are an extremely helpful resource when it comes to practising outputting the information you have learnt. If you have already found out what you will be examined on (see Section 14.1.1) you will know what the format and scope of the exam will be. Assuming there are no major changes to an exam's format or scope for that particular year, you can assume that the past papers will be representative and so be a good test of your learning so far. There are a number of ways you can use past papers, including:

- practise analysing questions and setting up plans to answers in order to test your recall and ability to adapt material to the set question;

- practise writing full answers in the time allowed;
- review your answers by checking them against your notes and highlighting any missed or inaccurate information;
- ask one of your lecturers to read your essay and comment on it;
- analyse your answer to see how it could have been improved (Did it need more information? Did it have a logical structure? Was your expression clear? Did you stick to the question?).

Your precise use of past papers will depend, in part, on the format of the exam; multiple choice, short answer, or essay. We will deal with this in more detail in the next chapter.

* Chapter summary

We said at the beginning of this chapter that effective revision makes the material you are revising easier to understand, remember, and apply. Always make sure that you:

- get yourself organized in your revision;
- set a clear timeline leading up to the exam;
- make your revision an active, rather than passive;
- don't lose sight of the big picture in terms of how detail fits into an overall structure;
- practice outputting the information you have been learning to ensure you can apply it to specific exam questions;
- answer past examination papers.

Chapter 15

Getting the most out of exams

➲ Introduction

We ended the last chapter on *Getting the most out of revision* by addressing the use of question banks and past papers as a means of practising how to output the information you have learnt. This final chapter on *Getting the most out of exams* will address the skills you need for the exams themselves. However, exam skills aren't just needed during exams; there are some important exam skills you need to develop for the day or so leading up to exams too. With this in mind, this chapter will begin by addressing what you need to do immediately before your exams, and then progress to give guidance about the exams themselves.

Remember: unlike most other topics covered earlier, you won't have this book to hand for guidance when you're actually experiencing the topic of the chapter – that is, when you are sitting in the exam room. So it is worth spending time reading carefully to make sure you remember all the key points, so that you have them in mind and ready to put into action, when exam time comes.

15.1 **Just before an exam**

If you have got yourself organized and managed to do a reasonable job of your revision, as described in the previous chapter, then there will be less of a temptation to do any last-minute cramming. You need to be realistic at this point though: there will always be more you could do, but at some point you need to stop. It is very unusual, even for students who have managed to get themselves very organized, to be able to walk into an exam room feeling completely prepared (after all, exams are – by their very nature – an unknown quantity; the only way to be 100 per cent prepared would be to know the questions before the exam!). It is therefore common to want to try and cram in a lot of revision just before an exam (either because your revision hasn't gone as well as you had hoped or because you are anxious), but if you do find yourself in this situation then be careful! There is work that is helpful to do at the last minute but, equally, some that can be unhelpful.

15.1.1 **The wrong sort of last-minute work**

Last-minute cramming is the wrong sort of last-minute work. There are a number of reasons for this.

Last-minute cramming is ineffective

If there is material that you still haven't covered the night before the exam you are unlikely to gain much benefit by covering it at this late stage. Your time would be much better spent going over things that you are already familiar with (see Section 15.1.2).

Last-minute cramming makes you forget other stuff

Last-minute cramming is often counterproductive because trying to cram in a lot of information (particularly new information) at the last minute can cloud your memory of what you already know and so has the effect of pushing out other information that you may have otherwise been able to retain.

Last-minute cramming makes you tired

Exams are hard work and require a lot of energy and concentration. Therefore, making yourself tired by staying up late the night before (or worse, through the night) is not good preparation. Your time would be much better spent trying to get some rest so that you are alert for the exam and so can make a good attempt at answering the questions, rather than being so tired you can't think clearly.

Last-minute cramming makes you worry

Last-minute cramming tends to make you focus on what you don't know rather than what you do know. Thinking about all the things you don't know the night before an exam is a bad idea! It makes you worry, it gets you stressed, it makes it difficult for you to get to sleep, which makes you more worried and more stressed... and so it goes on.

15.1.2 **The right sort of last-minute work**

While last-minute cramming is the wrong sort of last-minute work, there *is* work you can do just before an exam which is helpful.

Plan your use of time

It is important to plan how you will use the time you have in the exam *before* you get to the exam room. This doesn't have to be a last-minute exercise, but if you haven't done it as part of your revision strategy, now is the time to do it. You need several pieces of information to plan your use of time effectively:

- how long the exam lasts;
- how many questions the paper contains;

- how many of the questions you are required to answer;
- whether all the questions are worth the same number of marks.

Once you know this information you can decide how much time you will spend on each question, remembering to allocate time for choosing questions (if you have a choice), analysing questions, planning answers, and reviewing answers, as well as (of course) time to write your answers. For example, if you had a one-hour short-answer paper containing nine questions, each of which was worth 10 marks, you might work out how to allocate your use of time as follows:

- 1 hour = 60 minutes;
- 5 minutes per question (including analysing and planning time) = 45 minutes;
- which leaves 15 minutes checking time;
- 15 minutes checking for 9 questions = just over 1.5 minutes checking time per question.

If you had a three-hour essay paper containing 10 questions, all worth 25 marks, of which you had to answer three, you might allocate your use of time as shown in Figure 15.1.

Planning your use of time before the exam will give you one less thing to think about in the exam itself. It will also ensure that you at least attempt all the questions you are required to attempt, rather than simply focusing on the ones you are confident on. In an essay paper where you are required to write three essays, for example, you will usually get more marks by having a go at all three essays rather than spending time trying to make two of them really good ones. Even if you are not confident about the third essay and

FIGURE 15.1 Planning your use of time.

Time (minutes)	Task
5	Read questions
5	Choose questions
50	Question 1 – analysing (5) – planning (5) – writing (30) – reviewing (10)
50	Question 2 – analysing (5) – planning (5) – writing (30) – reviewing (10)
50	Question 3 – analysing (5) – planning (5) – writing (30) – reviewing (10)
20	Checking

think you can only write a fairly poor answer, the chances are your time would still be better spent attempting an answer rather than refining an existing one.

The night before the exam

We have already noted that the night before an exam is not a good time for learning new information. It is, however, a good time for practising your exam technique or checking your knowledge of what you have already learnt. So use your question banks and past papers to go over the information you have already covered, making sure that if you come across something that you don't know you look it up so you do know it. Going over pre-revised material in this way the night before and exam will both boost your confidence regarding how much you do know (you might be surprised!) and act as a reminder of the things that you had learnt but couldn't recall in response to a question.

The morning of the exam

The morning of the exam is obviously also not a good time for learning new information. As we have seen, cramming makes you forget other stuff; in particular it can cloud your ability to remember overall concepts. Instead, focus on reviewing main points to give you an overview of what you have learnt. If your exam is in the morning, it would be a good idea to not do any sort of revision on the morning of the exam at all.

15.1.3 Other last-minute preparations

Good exam technique is about more than simply learning information and then outputting that information in response to specific questions. Your preparation could have gone brilliantly, but if you are so tired you can't think clearly, or you turn up at the wrong location, your preparations will have been wasted. So here are a few more last-minute preparations to consider.

Get your stuff together

Getting everything you will need for the exam organized the night before will give you one less thing to think about on the day itself. You will need a pen you can write with comfortably for a long time, and at least one spare. Depending on the exam you may also need other equipment, such as a calculator or molecular models. Also think about what you don't need, which is anything that might cause your invigilator to suspect you of cheating if they found it on you. This includes mobile phones, MP3 players, scraps of paper with notes on, and scribbles on the back of your hand.

Double check when and where you need to be

Another thing you don't want to be worrying about on the morning of the exam is when and where the exam will take place. Make sure you have the latest information on the date, time, and location of your exam (sometimes these change due to timetabling clashes, which is why the *latest* information is important). Then double-check this with people who are taking the same exam, just to make sure. If your exam is taking place in a location you are not familiar with, it is worth either going to the venue beforehand or arranging to go with someone on the day who is sure they know where it is. Also, set your alarm to make sure you wake up in time, and then set another alarm, just in case. You may also want to let the people you live with know when you need to be up, so they can help make sure you're up too. Having more than one wakeup call reduces the chances of you worrying about not waking up in time and so makes it more likely that you will get to sleep. Plan to arrive at the exam room in plenty of time as racing in at the last minute (or late) flustered and out of breath is not a good way to start.

Get some sleep

We have already identified that last-minute cramming isn't helpful, but a good night's rest the night before the exam is. However, this is easier said than done, so you need to have a strategy. Firstly, make sure that you don't drink too much caffeine too late, as clearly this will keep you awake. Secondly, it is very unlikely that if you stop working at 11.00 p.m. and then roll straight into bed you will fall asleep quickly, and the longer it takes you to fall asleep the more worried you will become about not being able to sleep and so the more awake you will become.

To help you relax and sleep well you need to plan something relaxing to do in between stopping work and going to bed. It could be chatting to your housemates, going for a walk round the block, or watching a bit of television (as long as you don't start a whole film!). Anything that will help stop your mind racing and calm you down a little will make it more likely that you will be able to sleep.

Manage your stress levels

We have deliberately given this section the title 'Manage your stress levels', rather than 'Get rid of stress' because, first, it's unlikely that you will be able to get rid of stress altogether, and, secondly, some stress is helpful: it is your body's way of preparing itself to cope, making you more alert and attentive. All that we have covered so far in this chapter, and the previous chapter on revision skills, will help you to manage your stress levels, because one of the most stressful things, as far as exams are concerned, is lack of preparation. So getting organized will really help: the most stressful exams are always the ones you haven't got organized for. If you still feel anxious though, it can also be useful to talk to someone about your worries concerning exams. This could be a friend, your tutor, or someone in your university's student support services. Additionally, on the day of the exam itself, there are a number of things that will help you manage your stress levels:

- try to avoid talking to other students if you think if will increase your anxiety levels;
- get plenty of fresh air – exam rooms are often stuffy places, so get plenty of fresh air beforehand (this will also help you avoid talking to other students), and breathe deeply to try and calm your nerves;
- make sure you use the toilet before the exam – you really don't want to be needing to go half-way through the exam: it will be a major distraction, and will make you feel unsettled (and, therefore, stressed).

Exams are physically tiring. So there are also a couple of things not directly related to the management of stress that you should keep in mind to make yourself as prepared as possible from a physical point of view:

- try to eat breakfast – not everyone finds this easy, but exams are hard work and you need energy. At the very least, have something simple but high energy, like a banana.
- drink water – to keep you hydrated and help you think clearly (and take a bottle of water in with you).

15.2 Arriving in the exam room

When you arrive in the exam room there are a number of things you can do to get yourself off to a good start.

15.2.1 Get settled

Exams have a lot of build up and there is often a lot depending on them. So it is inevitable that you will be feeling at least a little anxious. Just by sitting down, taking a few deep breaths, getting out your pens, your clock or watch, and your bottle of water, and generally getting yourself comfortable in your chair will help you settle down.

15.2.2 Read the instructions very carefully

Once you have been told by an invigilator that you can turn over the paper, turn it over. The first thing to check is that it is the paper you were expecting (if it isn't, put your hand up – often exam rooms contain several different exams at once; you could just be at the wrong desk). Once you are sure you have the right paper begin to read the instructions carefully. The instructions should be familiar to you from your preparation, but check in particular the number of questions, how many questions you are required to answer, and how many marks each question is worth.

15.2.3 Remind yourself of your time plan

You should have planned your use of time beforehand (see Section 15.1.2), so this should be simply a case of reminding yourself what you had decided to do (but if you haven't – do it now). You may find it helpful to make a note of your time allocation on the question paper so you don't forget or become confused about what you had planned to do (which is easily done when you are under pressure).

15.3 Answering the questions

The areas we have covered so far in the sections on *Just before an exam* and *Arriving in the exam room* apply equally to all types of written exam, whether multiple-choice papers, short-answer papers, or essay papers. However, when it comes to actually answering the questions there are some exam techniques that are particular to different types of written exam. We will deal first with some general advice applicable to all question types and then address how these apply specifically to different types of exam paper.

15.3.1 General advice for all question types

Regardless of the type of exam paper you are faced with it's a good idea to have a clear sequence of steps in mind to help you answer the questions to the best of your ability. We suggest the following:

- analyse the questions;
- plan your answers;
- write your answers;
- review your answers.

Following these four steps will help ensure that you have a measured approach to what can otherwise be quite a panicky situation.

Many examination questions are structured so that there is an easy 'starter' part of the questions first, with the question building in difficulty. So, if you are struggling, it may be more beneficial to tackle the first part of each question rather than spending a long time on one full question.

15.3.2 Multiple-choice papers

Of the question types we have identified, multiple-choice papers require the least amount of analysis and planning. However, that's not to say you can dispense with analysis and planning altogether, you just need to approach it in a slightly different, and briefer, sort of a way.

Analyse the questions

Usually with multiple-choice papers you will not have a choice of questions you can answer. If this is the case then there is no need to read through the entire paper before you begin. Instead, simply read the instructions carefully (as described in Section 15.2.2) and make sure that the structure of the paper is what you expected. Once you have confirmed this you can begin to work on the individual questions. With multiple-choice papers simply start with question one and then work through the questions in sequence. When you come across a question you don't know the answer to, don't dwell on it too long, just move on to the next one and come back to the ones you are not sure about at the end.

Plan your answers

You neither have, nor need, much time to plan your answers for multiple-choice papers. However, a few moments spent planning will help make sure you don't rush headlong into a wrong answer. A good approach is as follows:

- read the question carefully;
- don't look at the possible alternatives yet – try to answer the question independently of the options given;
- look down the list of options to see if your answer is there;
- read all the other alternatives just to make sure.

Write your answers

For multiple-choice questions this is simply a case of ringing a response or ticking a box. However, make sure that you have read the question carefully. Sometimes multiple-choice papers include multiple-response questions as well as multiple-choice questions, that is to say questions that require more than one answer. For example:

Which of the following statements about the transition metals is true?

a) The transition metals typically have a complete shell of d orbitals. ☐

b) The transition metals exhibit a range of stable oxidation states. ☐

c) The transition metals have low densities. ☐

d) Transition-metal compounds are typically colourless. ☐

In this question, as indicated by the stem: 'Which ….*is* true?', there is only one correct answer (b).

In some questions, there may be more than one true statement. These are indicated in two ways. For example:

Which of the following statements about the transition metals are true (tick all that apply)?...

Or some take a more complicated form as follows:

Which of the following contains no unpaired electrons:

$[ScF_6]^{3-}$ $[CrF_6]^{4-}$ $[Mn(CN)_6]^{4-}$ $[FeF_6]^{4-}$ $[Fe(CN)_6]^{4-}$

A.	$[ScF_6]^{3-}$ only
B.	$[CrF_6]^{3-}$ and $[FeF_6]^{4-}$ only
C.	$[Mn(CN)_6]^{4-}$ and $[Fe(CN)_6]^{4-}$ only
D.	$[ScF_6]^{3-}$ and $[Fe(CN)_6]^{4-}$ only
E.	All of them

This second type of question is quite complicated and you need to work through each example, marking each one that you think contains no unpaired electrons and then seeing how your list matches with the options given.

Also, make sure you have understood the sense of the question. For example, check whether the question is asking you to indicate which is true or which is *not* true.

Your exam questions may also be set out as a grid. Here, you will be given a series of questions, the answer or answers for each will be found within the grid. An example of such a grid is given below:

State which species A–I in the grid answer the following questions. Each may be used more than once or not at all.

A $[Mo(PF_3)_3(CO)_3]$	B $[RhHCl_2(PPh_3)_3]$	C $[Ni(\eta^5\text{-}C_5H_5)_2]$
D $[IrCl(CO)(PPh_3)_2]$	E $[Rh(\eta^5\text{-}C_5H_5)_2]$	F $[V(CO)_5]$
G $[Cr(\eta^6\text{-}C_6H_6)_2]$	H $[Ru(\eta^5\text{-}C_5H_5)_2]$	I $[Mo(PMe_3)_3(CO)_3]$

i) Which one or more metallocenes would be stable?

ii) Which molecule contains the phosphine with the smallest Tolman cone angle?

iii) In which molecule/s is the metal in a zero formal oxidation state?

iv) Which compound/s can undergo oxidative addition reactions?

v) Which compound/s could be prepared by an oxidative addition reaction?

vi) Which of A or I has the lowest C–O infrared stretching frequency?

vii) Which one or more species forms anions readily?

Review your answers

First, check that you have answered all the questions, and then review your answers, starting with the ones you were least sure about. If you amend an answer make your change very clear; there will be advice on how to do this in the instructions at the beginning of the paper.

15.3.3 Short-answer papers

Short-answer papers can also be tackled using the same four steps. Again, one of the keys to doing well on these papers is being strict with your timing – don't spend a lot of time thinking about each question. If you know the answer, answer it quickly, if you don't know the answer, move on and come back when at the end, when you may have time to think about it more carefully.

Analyse the questions

As with multiple-choice papers, short-answer papers often don't give you a choice of which questions you can answer. Again, if this is the case you don't have to read through the entire paper before you begin. Start with the first question and read it carefully to make sure you understand it, perhaps underlining key words to help you focus on the exact meaning. If after a quick reading you are not sure how to answer it, move on to the next question and come back to the ones you are less sure of at the end.

Short-answer questions usually require specific factual information for the answer, and not detailed description. Usually the number of marks to be awarded is given on the exam paper, so that can be a guide to the amount of information you need to give. Typical short-answer questions might take the following forms.

1. Describe and explain the chelate effect. (4 marks)
2. Explain why $[Mn(OH_2)_6]^{2+}$ is pale pink, whereas $[MnO_4]^-$ is intense purple in colour. (4 marks)
3. Calculate the spin-only magnetic moment for $[Cr(OH_2)_6]^{3+}$. (4 marks)
4. Sketch the isomers of $[Pt(NH_3)_2Cl_2]$. (4 marks)

Plan your answers

With short-answer questions, in briefly planning your answers you are trying to make sure that you select information relevant to the question and put it into a logical and coherent order. Thus, in each of the questions above, the marker is looking for very specific factual information. To answer the question, you will need to note down a few keywords to stimulate your memory and organise your thoughts.

Write your answers

Use your brief plan to prompt you of your points and structure your answer. You only have a short time to write each question, so the important point is to make sure that all the key facts are included. Thus, the diagrams do not need to be a great work of art, but you must remember to include key details. When writing your answers, brevity and factual content are the keys to a good answer.

Review your answers

Check that you have answered every question that you needed to answer (probably all of them), then review your answers. As with multiple-choice papers, start with the questions you are least sure about because these are probably the ones you will be able to make the most improvement to.

15.3.4 Essay papers

Often with essay papers you will have a choice as to which questions you can answer. If this is the case then make sure you read through the all the questions carefully before you choose which one (or, more likely, ones) you are going to answer. As you read the questions, mark the ones that you think you will be able to answer the best, then go back and read them again to make a final choice. As with multiple-choice and short-answer questions, start with the questions you are most confident with to get you off to a good start and stimulate your thinking.

Analyse the questions

Analysing questions is particularly important for essay questions. With multiple-choice and short-answer papers your answers will usually just be facts or isolated pieces of information. With essay questions, however, there is more work to do to at the analysis stage in order to be able to organize your answers logically and coherently. As we said in Chapter 14, *Getting the most out of revision,* essay questions will usually be very specific. So, instead of asking 'Please tell us everything you know about high-temperature superconductivity', they are much more likely to ask you to, for instance, 'Discuss the origins and applications of high-temperature superconductivity. Illustrate your answer with relevant examples.' To apply your knowledge to this particular question requires you to analyse the question briefly, using the same technique that we identified in the section on *Analyse the question or brief* in Chapter 7, *Writing essays and assignments.* The key information to identify in the question or title is:

- the subject of the question;
- the instruction;
- the key aspect;
- other significant words.

Plan your answers

Planning answers is particularly important for essay questions because it helps you to:

- select information that is relevant to the question;
- put this information into an order that is logical and coherent;

- write your ideas down at an early stage to help you remember your key points;
- monitor how much information you are covering in the time allowed.

Write your answers

When answering essays under exam conditions it is important that you keep to your time plan as it can be easy to lose track of time if you get into the flow of writing. One way to do this is to make a point of checking the time when you come to the end of a section in your answer. This is preferable to checking the time at random intervals (which could be insufficient) or too frequently (which could distract you from writing fluently). Also, keep looking at the question to remind yourself of the focus of your answer, and keep looking at your plan to remind yourself of what you need to cover. Again, the end of each section of your answer makes a good natural break for checking you are still on track.

Review your answers

Being able to review your answers is where the benefits of having planned and monitored your use of time become obvious. Not only should you have had enough time to make a reasonable attempt at each of your chosen questions, you should also have time left at the end to go back and review your responses. As we highlighted in Chapter 7, *Writing essays and assignments*, making time to review your work can make a significant difference to your final grade.

As with multiple-choice and short-answer questions, first check that you have answered every question that you chose to answer, then review your answers, starting with the one you were least sure about. The purpose of reviewing answers in an exam context is slightly different from reviewing answers in a coursework setting: you don't need to finish up with a very polished piece of writing that is spelling error-free and grammatically faultless, just one that answers the question well. Of course, if you do end up with a polished piece of writing then that's good, but it's not the primary purpose of a review in an exam context.

Instead, you should be looking at the bigger issues, such as:

- Have you answered the question?
- Does the order of your material make sense?
- Is the meaning of what you've written clear?
- Have you made appropriate use of figures to supplement what you've written?

These are the main issues; you will get more marks for a response that has spelling and grammatical errors but answers the question well than you will for a response that has perfect spelling and grammar but doesn't answer the question well.

If you need to make significant structural changes to your answer, such as moving a paragraph from one section to another, you need to do this in such a way so as to not confuse the people marking it. In a word-processed script this would be straightforward, but in a hand-written script you will need to annotate it in a way that makes it clear you are moving a section rather than, for instance, deleting it.

* Chapter summary

Performing well in exams requires particular skills. Whilst your exam performance will be heavily influenced by how well you have revised, it is how you conduct yourself during an exam itself that will ultimately determine your grade. What you do just before an exam and when you arrive in the exam room are important stages of your preparation. Different types of question types require slightly different techniques but in order to succeed you should:

- read the instructions carefully;
- analyse the questions;
- plan answers to the questions;
- answer easy questions first;
- keep an eye on the time, do not spend more than the allotted time on a question;
- review your answers.

Chapter 16

Using feedback

Introduction

So, the assignment you worked so hard on has been marked and returned to you. You take a quick look at the mark and that's quite good, so you put the assignment down among the pile of paperwork on your desk and it doesn't get looked at again. You feel quite happy but you have also just denied yourself a valuable learning opportunity: you got quite a good mark but why wasn't it really good? What could you have done better? How can you get a really good mark next time?

This is where feedback, and how you use it, is important, and is a vital component in the foundational skill of developing yourself that we identified in the first chapter. It is all too easy to make excuses for not doing anything: 'I'm too busy at the moment, but will look at it when I get the chance'; 'I did the piece of work so long ago, there's no point in going back over it now'; 'We've finished that module, so there won't be anything useful anyway'; 'I never understand what Dr Blogs writes, so there's no point in looking at it'. No doubt you can come up with quite a few more seemingly plausible reasons for not bothering to revisit the piece of work, maybe not even bothering to collect it in the first place. However, whatever form the feedback takes, even if it is really limited and you only have the bare marks, you can still make use of it to improve for next time.

16.1 What is feedback?

Ideally, feedback should serve three key functions:

1. provide you with a clear indication of how well you are progressing in your work in relation to the standards established by your university;
2. give guidance as to how you can improve for when you come to prepare the next piece of work;
3. flag up things you are doing well and should therefore continue doing.

At its most basic level, then, feedback takes the form of the marks you get for the piece of work: these marks give you an overall indication of how well you are doing. In terms of specific guidance, marks on their own are clearly not very helpful; nonetheless, you can still use them to help you improve (see *How to make use of feedback*, below).

One level up from this in terms of usefulness is negative feedback that indicates what you got wrong or did badly. This is probably the most common form of feedback and it may take the form of specific comments such as: '...you did not discuss the role of…' or '... this is wrong'. The marker may also include more generic comments such as '…the essay was poorly structured' or '...not enough detail'. These types of comment are very common, in part because they are usually the easiest type of comment for a marker to make. By pointing out what is wrong, the marker is providing limited guidance as to areas for improvement, but it is often not clear exactly what you should do to improve, particularly in the case of the generic comments that may be quite vague. In this context, therefore, the best negative feedback is that which identifies the error and then provides guidance as to how to overcome it. This may take the form of full examples, or perhaps pointing you in the direction of specific resources that can help you:

'Make sure that you use the appropriate referencing style for your essays – look at the section on referencing in the course study guide.'

'You have misunderstood the mechanism of the reaction – see pages 32–35 in your organic textbook.'

Another common form of feedback is positive feedback. Here the marker not only provides encouragement: '...this was a good essay…' but also identifies specific aspects of the work that were well done:

'This was a very good essay and I particularly liked your use of diagrams to support the arguments you were making.'

As with the negative feedback, the value of positive feedback is greatly increased where examples are given so that you are clear about what things you should carry on doing. Inevitably, the quality and level of detail of the feedback you receive will vary. Indeed, recent National Student Surveys have all identified feedback as the aspect of higher education that students find less satisfactory. However, it must be remembered that feedback is a two-way process and you will only benefit from it if you engage with the process. In this way, even if the feedback you receive is not always very detailed or is late in coming, you will still be able to identify ways of improving what you do.

Try this: Identifying feedback

Make a list of all the possible ways that you might receive written and non-written feedback from your tutors.

16.2 When do you get feedback?

Many students only think of feedback in terms of written comments on coursework assignments. Indeed, this is the most obvious and one of the most common forms of feedback. However, feedback comes in a variety of forms and unless you are ready to recognize when you are receiving feedback, you will be missing out on valuable learning opportunities. Let's now consider the range of different feedback types that may be available to you.

16.2.1 Written comments on written assignments

As stated above, written comments on written assignments constitute the most obvious form of feedback and are probably the most *common* form of formal feedback. These types of comments usually take two forms.

1. Generic comments, usually written at the end of the piece of work, or on a separate cover sheet. These comments tend to relate to the overall style and quality of the essay.
2. Specific comments that usually take the form of annotations made on the script itself. These may relate to stylistic issues, such as paragraph structure or referencing, or they may be subject specific, such as the identification of a factual error or the omission of a point in the argument.

16.2.2 Comments on oral presentations

These feedback comments may be given in the form of written comments on a mark sheet, but they are often given just as verbal feedback following the presentation. Such verbal feedback can be very useful because it is the most immediate form of feedback, being given within a few minutes of the delivery of the piece of work. However, verbal feedback is also transient: if you don't write the comments down you are likely to forget them very quickly and then you won't derive any long-term benefit from them.

16.2.3 Comments during tutorials

Tutorials are often a source of very useful discussion about specific topics in the curriculum (see Chapter 5, *Making the most of tutorials and group work*). However, as with the verbal feedback on oral presentations, the points made by a tutor will be transient and quickly forgotten unless you note them down. Tutors probably consider that they are providing you with valuable feedback during tutorials but students often do not recognize it as such.

You may have sessions with your personal tutor specifically to discuss your progress as part of personal development planning (*PDP and career planning*, Chapter 17).

During these sessions you should be able to reflect on your own progress and receive feedback on your progress from your tutor. These sessions should enable you to use feedback to 'feed forward' to future development and improved performance.

16.2.4 Comments during practical classes

Practical classes offer another very useful way in which students can engage in discussion with tutors or demonstrators. As a consequence, they can be a rich source of feedback for example in relation to experimental technique, analysis and presentation of results, or in understanding of the topic itself. Again, however, much of this feedback is transient and it is very important that you note the comments before they are forgotten.

16.3 How to make use of feedback

If you are going to benefit from feedback that is provided, then you must take steps to engage with it as an active process. This doesn't just mean looking briefly at the comments. Engagement requires a detailed reflection of how the comments relate to the piece of work you have submitted and consideration of how you should change your approach to future pieces of work.

There is often a perception that if feedback is not provided very quickly, then it is of no value. The usual reasons for taking this view are:

- that it is several weeks since you did a piece of work and you can no longer remember much about it;
- that the module for which you did the piece of work has now finished and therefore the feedback no longer appears relevant.

Both these reasons, while initially understandable, are false. There are always useful pointers that you can take from feedback, though it may be that sometimes it requires more work on your part to extract the maximum value.

For the most part in this section, we will focus on making use of written feedback since this is the most common form of formal feedback on coursework; however, much of the guidance also holds true for verbal feedback. The difference between the two is that in the latter case, it is up to you to make sure that you make a record of the feedback.

16.3.1 Engaging with written feedback

If you have a cover sheet, as is used by many universities, or if you have been given summary of written comments at the end of the essay, then read these carefully before doing anything else. Divide the comments into two groups.

1. Generic comments: ideally these take the form of 'feed forward' comments such as identification of broad areas for improvement. This could include how to improve your overall writing style, to reference better, or to draw graphs and present data better. There may also be positive comments, identifying things that you have done particularly well and should carry on doing in future work.
2. Subject specific comments: these will normally be related to factual information such as the identification of factual errors or omissions. Again, markers should also flag up good points, so you get a feel for what you are doing well as well as what needs improving.

Having categorized the comments, read through the piece of work carefully; after all it may be some time since you wrote it. As you read through, bear in mind the comments that were made and try to be objective in your evaluation. Also, make sure that you can identify what aspects of the essay or report the comments relate to. You should also note any annotations on the script. With the generic comments, make sure you understand how the comments apply to your work and how you can improve what you have done. Make a note for future reference so that you can carry them forward to your next assignment: although the next assignment may well be for a different module, generic issues such as referencing style will still be relevant. Likewise, make a note of aspects of your work that were highlighted as being particularly good. With subject-specific comments, check that you understand any factual errors or omissions. Make a note to go alongside your lecture notes to make sure that the point is clear for revision purposes.

Look at your marks, and check them against any marking criteria that you might have been given for the assignment or against any generic criteria that your department might have published. Make sure that you appreciate how your work fits against the criteria and why it falls into the specific mark band that was allocated. Again, if you don't understand, then ask.

For example, if you obtained a mark of 55%, the marking scheme might state that the work has the following characteristics.

> The essay is well organized, displaying understanding of the main issues. There may be a few, minor errors or poorly expressed ideas. There is a significant dependence on lecture notes and/or textbook material.

As you read through your essay, make sure you can identify why the marker placed it in this category: were there some minor errors? Did you mainly use lecture notes or textbooks rather than research literature (see Chapter 5, *Working with different information sources*)? Think about how you could do it better next time.

Feedback may not always be clear! For example the comments may be illegible or perhaps you don't understand how to improve even though the marker has highlighted a weakness. You might not understand what the marker means by some comments. If you are not sure what to do, then do ask your tutor to explain in more detail how you can improve.

Make sure that you take the feedback forward and use it to improve your next piece of work and your understanding of the subject. When you have taken the points forward, also check the feedback on the next piece of work to ensure that you have indeed improved.

> **Try this: Using feedback to develop skills**
>
> Try this feedback activity with a friend. Swap the feedback that you each receive on your next written assignment. Identify the skills that your friend needs to improve on and encourage them to write a brief account of how they will improve them. Ask your friend to identify the skills that you need to develop and then you write a brief account of how you will improve them.

16.3.2 Engaging with oral feedback

As indicated earlier, this may take various forms, for example in the laboratory, a tutorial, or as feedback on a presentation. The danger is that this form of feedback is very transient and easily forgotten therefore you must be sure to take notes so that you can remember the specific points. The principles, though, are the same as for written feedback: divide the feedback up into generic and specific aspects, and make sure that you understand what the issues are and how to take them forward.

Feedback, then, is one of the keystones of good learning, but you can't treat it as a passive process, otherwise you will not progress and you will make the same mistakes again and again. Engagement is all important.

* Chapter summary

The most important aspect of feedback is that you need to think about what it means, see how it relates to the work that you have had marked and then to see how you can use it in future to improve your performance. You should always try to:

- read through the comments and categorize as generic or subject specific;
- review your work, relating the marker's comments to the work;
- identify feedback that you can carry forward to your next piece of work;
- look at the marking criteria to understand why you got the mark you did;
- identify any feedback or comments you don't understand;
- check that changes you have made have been effective in future work;
- if in doubt, ask!

Chapter 17

PDP and career planning

Introduction

Personal development planning (PDP) is a structured process through which undergraduates are encouraged to reflect on their own learning and to plan for personal, educational, and career development. The aim of personal development planning is to familiarize students with strategies that encourage personal responsibility for lifelong learning. It should help you to gather the evidence that you will need to ultimately prepare an excellent curriculum vitae and draw on useful examples during an interview. If you make a success of PDP at university you will find it much easier to identify your learning and training needs as you move through your own career. In this chapter we look at how you can use PDP to identify your skills and qualities, plan to develop them further, and then to use evidence gathered during this process to produce an excellent curriculum vitae.

17.1 What is PDP?

PDP is often the collective term used to describe two separate things; the process of reflecting on learning and performance and planning for development and the process of recording that reflection, possibly with the provision of supporting evidence. Obviously, it is possible to carry out the reflection and planning without documenting the process, but most intuitions will require the completion of some form of portfolio or e-portfolio where you will record both the process and supporting evidence.

PDP is often associated with the development of transferable skills but the process should cover your academic achievements too and should draw on your experiences within the university, the world of work, and any extracurricular activities. The process of PDP should be supported by your institution, probably through your department. For example, you may be supported in your reflection through:

- meetings with your personal supervisor where you are encouraged to reflect on your progress so far—the meeting may have a formal agenda, be organized around providing feedback on a piece of work, or may be less structured;

- questions on module evaluation questionnaires that ask you to reflect of your progress;
- skills development workshops where you provide or receive feedback on transferable skills, for example, on oral presentations;
- 'badging' or clear identification of activities, handouts, etc. that are part of the PDP process;
- any other activity where you are encouraged to consider how you have performed in an assessed or unassessed activity.

It may be that PDP is closely integrated into your programme and little reference is made to it. In this case you should still be aware that those activities through which you reflect on your progress are crucial to developing lifelong learning and to future career development.

It is highly likely that your institution or department will define how you record your reflection and any associated evidence. This is usually by the use of a portfolio of evidence, or an electronic version called an e-portfolio. The portfolio (or e-portfolio) could be a free-standing document or piece of software or it could be integrated into your institutional virtual learning environment or portal/internet. Whatever its form you will be provided with guidance on completing it.

Whatever the support mechanisms that are put in place to help you engage with PDP ultimately the responsibility for carrying out the reflection and completing your portfolio lies with you. You should be prepared to:

- be aware of opportunities provided by the university, both inside and outside the formal curriculum;
- make effective use of the resources provided to help you plan for your own educational and career development;
- increase your awareness of your skills, qualities and capabilities;
- set goals and plan actions to monitor, review, and develop your own progress;
- maintain evidence that demonstrates you capabilities;
- use the approved portfolio of similar system to effectively record and reflect upon your learning and achievements.

The PDP process can be broken down into a cycle of monitoring your own progress and achievements, recognizing strengths and areas for development, planning to improve skills, recording the process with evidence, monitoring new achievements, etc. This cycle is shown in Figure 17.1.

17.2 Building your portfolio

Your portfolio may have another name within your institution. In this chapter we use the term to describe wherever it is that you record your reflections on your progress and

FIGURE 17.1 The PDP cycle.

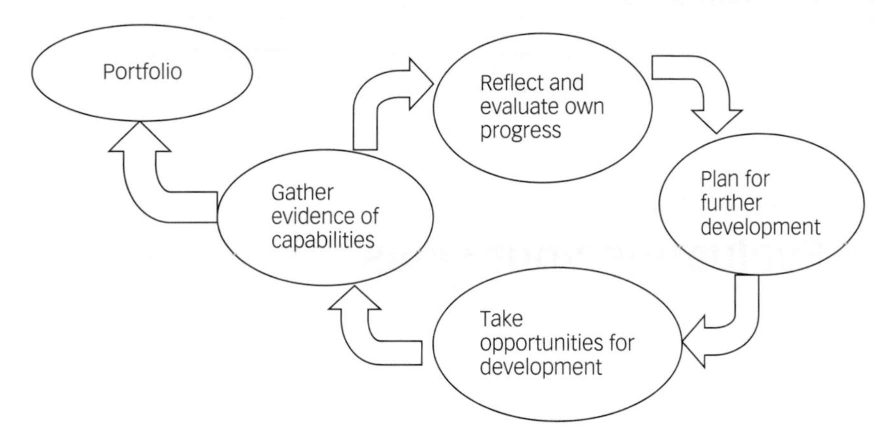

keep any evidence in support of your self-evaluation and planning. The portfolio may be hard copy or electronic. Your portfolio may be divided into various skills areas under which you record evidence. These skills may include group work, oral communication, written communication, numerical skills, IT skills, problem solving, learning to learn, organization and time management, working safely and professionally. It may be that each broad skill area could be broken down into individual skills. For example, the broad skill area of group work could be further subdivided into:

- contributing in labs, tutorials, seminars, and workshops;
- assuming a number of roles, e.g. team leader, team player;
- interacting with a range of people to obtain necessary information and assistance;
- giving and accepting constructive criticism;
- behaving towards colleagues with integrity and honesty.

If you are provided with a skills framework such as this for a number of broad skills then you are likely to be given advice on how to evaluate your performance against each one. You may be required to judge your ability in each skills area, perhaps rating yourself as novice through to expert, and to provide evidence to back up your evaluation.

Evidence can come from any aspect of your life; from university, extracurricular activities, such clubs, societies and volunteering work, as well as from paid work, social situations, and previous experience. The nature of the evidence can vary widely. For example you could use textual materials, such as feedback on written work and assignments, results from problem sheets or tests, laboratory and project reports, posters, and PowerPoint presentations. If you have an electronic portfolio, you could also be more creative and use photographs and videos, emails, blogs and wiki entries, and links to social-networking sites. You can also use personal communications from friends, employers, colleagues, etc.

Try this: Identifying evidence

You want to provide a piece of compelling evidence to demonstrate that you are a very effective team leader. What evidence could you include in your portfolio?

17.3 Evaluating your skills

When you apply for graduate employment opportunities you will either provide a comprehensive curriculum vitae or will complete an application form, probably online. You must assume that everyone applying for that post will have the appropriate qualifications. So having, say, a 2i MChem in Chemistry will certainly not be enough on its own to get you an interview. You must ask yourself what makes your application stand out from the crowd: employers are looking for well-qualified applicants but they are also looking for well-rounded individuals who can slot effectively into their organization and make an immediate contribution. What they are looking for are applicants with well-developed transferable or interpersonal skills. If you look at advertisements for graduate positions more than 60% do not specify a degree subject but they all describe desirable skills and qualities. It is essential that you not only develop these skills and qualities but that you recognize that you have done so and can provide examples. That is what PDP is ultimately leading you to. So, for successful PDP you need to evaluate your skills, identify where and how you can improve them, and, for potential employers, you need to be able to describe those skills and describe an instance where you used them effectively. These transferable skills are those that we have mentioned already; group work, numeracy, communication, IT, problem solving, etc. They can be drawn from any area of your life.

Activity

Read the following passage and decide what skills Titus had to use in order to resolve the problem.

Matthew, Beth, and Titus are doing a group project. Beth feels it's unfair that Matthew isn't pulling his weight. He expects the others to do all the work and won't turn up to scheduled meetings. Beth and Titus track Matthew down after a lecture and as the three of them try to discuss the issue the situation becomes heated and Beth and Matthew begin to argue. Titus is able to calm the situation by suggesting that although they had agreed to split all of the work three ways, clearly Matthew has not done his share so far. Titus suggests that as Matthew hasn't done any of the practical work he should prepare the poster presentation that is required, using the results that Titus and Beth have obtained. Titus asks Matthew if he agrees with the compromise and he agrees.

Some of the skills Titus showed he had in this situation were:

- keeping calm in difficult circumstances;
- effectively mediating;
- good listening skills;
- logical thinking;
- problem solving;
- being able to appreciate other peoples' views.

Try this: Identifying skills developed through problem solving

Ask each one of a group of friends to identify a problem they were faced with and managed to resolve. The problem doesn't just have to be connected to your course; it can be from any area of life. For example, it could be an issue with house mates, something connected to work experience or any club activities. It may be a problem that affected only you or affected others as well.

As each member of your group describes their problem, make a list of the skills that you think they demonstrated.

Group work can be broken down into several subskills. These subskills are often easier to evaluate yourself against than the broad skills areas. Evaluating yourself takes practice and honesty. Students often give themselves a lower rating with time as they gain more experience and are able to make a better informed judgement.

Try this: Auditing skills levels

For the following subskills associated with group work give yourself a rating and identify an opportunity for further development.

	novice				expert
Do you make a contribution in tutorials, seminars, workshops?	1	2	3	4	5
Can you assume a number of roles, e.g. team leader, team player?	1	2	3	4	5
Do you interact with a range of people to obtain information and assistance?	1	2	3	4	5
Can you give and accept constructive criticism?	1	2	3	4	5

What opportunities might you have in the next few months to practice and develop these skills?

17.4 Work experience and PDP

Many science degree programmes offer students the opportunity to carry out a period of work placement. This is usually for a whole academic year but can be for shorter periods of time. Such work experience is invaluable in developing those transferable skills that employers are looking for in graduates. Your PDP portfolio is a good way of capturing those experiences and reflecting on your personal and professional development during the placement. Indeed, completion of a portfolio or reflective log may form a major part of the assessment of the placement. During a work placement you will be expected to take responsibility for your time keeping and general organization. You will have to learn new areas of chemistry and related topics and you may be required to plan and carry out a project. Some of the skills that you will develop and provide evidence for include time management, team working, oral and written communication, independent learning, IT skills, experimental design, data interpretation, etc.

Even if you do not take part in a formal industrial placement scheme you may well be doing paid part-time work or voluntary work. These also provide valuable evidence of the development of skills that employers value and that should be logged in your portfolio to be used later in your CV.

Try this: Identifying skills from work experience

Alex spends the third year of his degree on placement in a multinational pharmaceutical company. He is working on a new formulation for sun screen and is part of a team of five chemists.

Ali has a summer job running a play scheme at a youth club. She organizes and delivers sports activities for twenty 12–13 year olds.

Make a list of the skills that Alex and Ali develop during their work experiences and suggest forms of evidence that they might gather for their portfolios.

17.5 Writing your CV

As we have already mentioned you should assume that everyone applying for any job that you apply for will have the appropriate qualifications. It is the other information that you present regarding your skills and qualities that will make you stand out from the crowd and make a recruiter want to meet you at an interview. So your curriculum vitae (CV) is your one opportunity to secure an interview. Your CV should make the most of those qualities and interpersonal skills that make you a well-rounded individual. This is where your PDP portfolio comes in to its own. You should have logged examples of activities where you have developed or effectively used a wide range of interpersonal skills. Use this information to remind yourself of those skills and to supply examples, either in your CV or later at an interview.

When it comes to presenting your CV there are many styles that are acceptable but there are some pieces of information that all CVs should include.

- Your personal details—name, address, contact telephone numbers and email address. You are not obliged to include a date of birth or marital status.
- Your educational history, in reverse date order—list your education from secondary school onwards with details of the institution and qualifications gained.
- Employment history, in reverse date order—include all your work experience, whether or not it is relevant for the position you are applying for. Give details of the employer and the role you had and identify any specific responsibilities, skills used.
- Other skills, such as a driving licence, specific IT skills, language skills, etc.
- Other personal achievements, such as awards and prizes, details of leadership roles, for example within clubs and societies, details of volunteering, in fact any other activity that enables you to demonstrate your personal skills.
- Some people like to include details of personal interests—be careful here. It might be relevant to mention a passion for learning languages or travel but less so if your passion is playing computer games or drinking with friends.
- The names of two referees, having checked with them beforehand that they don't mind you doing so. As a recent graduate it would be most appropriate to use at least one referee from your university, usually someone who knows you quite well. The best person is often your personal tutor or final-year project supervisor.

You might like to start your CV with a short personal statement that 'sums you up' before the detail. Something along the lines of '*I am an enthusiastic synthetic chemist with outstanding laboratory skills. I am a team player with excellent interpersonal skills and a desire to make a meaningful contribution to a major pharmaceutical company*'. If possible, try to target this statement to one or more of the key personal characteristics identified in the job advert.

Remember that all the information you present in your CV must be accurate. Do not be tempted to embellish the truth, as it could lose you the job if you are discovered. Make sure your CV is clear and concise. If you are unsure about the content or layout get some feedback from a friend, tutor, or careers advisor. Your CV should be well laid out, unfussy and easy to read. If it is at all unattractive or disorganized it will be placed on the 'reject' pile by the recruiter. Time spent planning your CV is a good investment as a document that looks as if no time or effort has been put into it is unlikely to prompt a positive response from an employer. You will need to update your CV regularly and to tailor it to each role that you apply for. For example, if you are applying for a position as a synthetic organic chemist you may want to highlight your laboratory skills and the topic of your final-year project. However, if you are applying for a position in the marketing department of a chemical company you may want to highlight your written communication and team-working skills. Carefully reading the job description and paying attention to the skills that the employer is looking for will help you know which areas of your CV to highlight. If you are applying for a position using a CV, always include a covering letter that indicates why you are interested in the position advertised and introduces relevant points in your CV.

It may be that the application process is through an application form rather than submission of a CV. In this case, the application form is your only opportunity to sell yourself and much of the same guidance applies as for using a CV. Make sure that you always read the form carefully several times before attempting to complete it. If the form is a paper copy rather than online, then keep the form neat and avoid crossed out text and spelling mistakes. If you are worried about how you will fit everything onto the form photocopy it first and then you can practice. ALWAYS fill in the section that asks you to say something about yourself and why you are suitable for the job. Leaving this blank means an initial selection process could be to discard those without anything positive or interesting to say. It looks as if you cannot be bothered to make the effort.

Try this: Applying for a graduate position

Figure 17.2 shows an advertisement for a graduate vacancy. Analyse the skills and qualities that might be required for the position and produce an appropriate CV and covering letter.

FIGURE 17.2 Advertisement for graduate vacancy.

Graduate Chemist

PaxoJones are a healthcare company at the cutting edge of pharmaceutical development. We have a large team of scientists at our research and production facilities at Whittlepool and are looking for an enthusiastic chemist to join our New Product Development Team who take new PaxoJones products from concept to market.

The Role

Based within the New Product Development Team you will work as part of a multidisciplinary team developing new product concept through the pilot plant stage to production, advising on formulation, packaging, and marketing. You will report to the Director of Development. You will present progress reports at monthly team meetings and liaise closely with colleagues in quality assurance, packaging, and marketing.

You should have, or be expecting to obtain, a good honours degree in chemistry/analytical chemistry. You should be self-motivated, dynamic, and able to work as part of a diverse team and under minimum supervision. If you feel you meet the criteria please email your CV and covering letter to:

Mrs P Potter, Head of Human Resources
PaxoJones PLC
p.potter@paxojones.com

17.6 Where to go for help

We hope that this chapter has been useful in helping you to think about the future and to start to plan for an effective CV. However, there are many more sources of help and support available to you. The first place that you should for help in preparing a CV, preparing for interviews or looking for ideas for future careers is your universities careers service. They will be able to give you sensible advice and they probably organize careers fairs. They may well have direct contact with many graduate employers. You should find out the name of your careers advisor. You may have an advisor who specializes in the sciences. Look at the careers centre's website. That may provide you with all the information that you need. If you do want to talk to someone find out whether they run a 'drop in' facility.

As you approach the end of your time at university you should have pretty clear ideas about what you want to do next. If this is not the case then you need to do some research. Here are a few questions to ask yourself:

- Are there any careers fairs coming up at your University?
- Have you spoken to your careers advisor?
- Does your possible career path affect your module choices or vice versa?
- Do you know what area or companies interest you?
- Does geographical area affect your choices?
- Do you know the closing dates for the graduate schemes you are interested in? Some of these are very early in the academic year?
- If you plan to do postgraduate study should you choose a relevant final year project?
- Many major companies recruit staff through recruitment companies. Temporary positions of this type can lead to permanent ones and even if they don't you can still gain valuable work experience.

Another useful place to look for careers advice is the Royal Society of Chemistry. Their website has links to their own careers service and also includes information on personal development, vacancies and working towards Chartered Chemist status (www.rsc.org).

You could also look at Prospects, the UK's official graduate careers website. This site has three main sections: careers advice, which gives you access to careers consultants, information what you can do with a specific degree and much more; jobs and work that provides advice on CVs, interviews, job hunting, a job search, to name but a few; and postgraduate study that profiles institutions and courses (http://www.prospects.ac.uk).

Milkround Online is useful for searching for internships and job vacancies, by industry and area, and includes upcoming closing dates for various graduate schemes. The site provides lots of useful advice, insights into working in a wide range of occupations, and details of forthcoming recruitment events and fairs (http://www.milkround.com).

* Chapter summary

In this chapter we have discussed the reasons for engaging with PDP, the benefits of reflecting on your own learning and achievements, and how the outcomes of PDP can be used to help you build an effective CV. PDP underpins all your learning at university and should enable you to gain the most from your time in higher education. In order to get the most out of it you should:

- be prepared to reflect on your achievements and development of interpersonal skills;
- record your achievements with supporting evidence;
- review your portfolio of evidence when planning your CV;
- ensure your CV is tailored to a particular position, is well structured, and well presented;
- make the most of the support offered by your careers service.

Index

V

W

PLATE 1: Colour wheel

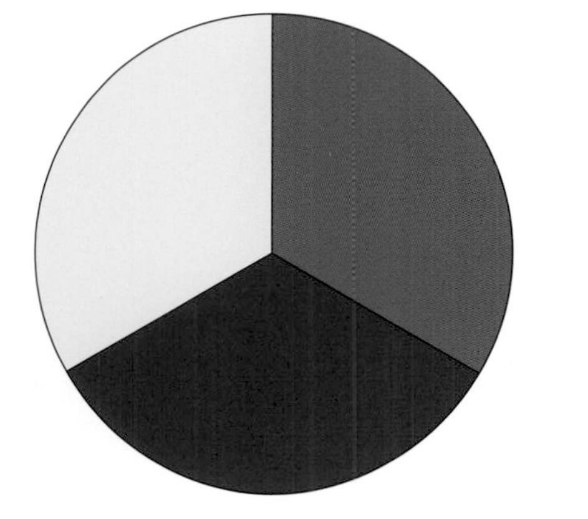

Primary colours

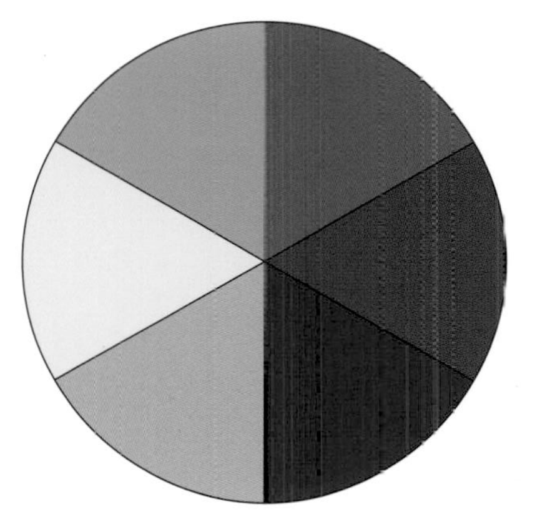

Secondary colours

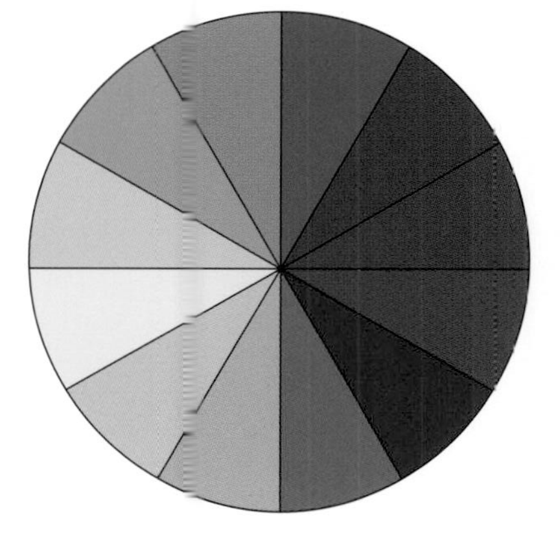

Tertiary colours